EUROPEAN DESERTS

EUROPEAN DESERTS

SANDY WILDERNESS ON A GREEN CONTINENT

CHRIS STONE

PELAGIC PUBLISHING

First published in 2026 by
Pelagic Publishing
20–22 Wenlock Road
London N1 7GU

www.pelagicpublishing.com

European Deserts: Sandy Wilderness on a Green Continent

Copyright © 2026 Chris Stone

Photographs copyright the author unless otherwise credited

The moral rights of the author have been asserted by him in accordance with the Copyright, Designs and Patents Act 1988.

All rights reserved. Apart from short excerpts for use in research or for reviews, no part of this document may be printed or reproduced, stored in a retrieval system, or transmitted in any form or by any means, electronic, mechanical, photocopying, recording or otherwise, now known or hereafter invented, without prior permission from the publisher.

https://doi.org/10.53061/SECA1952

A CIP record for this book is available from the British Library

ISBN 978-1-78427-520-4 Pbk
ISBN 978-1-78427-521-1 ePub
ISBN 978-1-78427-522-8 PDF

EU Authorised Representative: Easy Access System Europe – Mustamäe tee 50, 10621 Tallinn, Estonia, gpsr.requests@easproject.com

Cover image: Bardenas Reales, Spain © Getty Images / Inigo Cia

Designed and typeset by BBR Design, UK

5 4 3 2 1

Printed and bound in India by Replika Press Pvt. Ltd.

CONTENTS

Acknowledgements	vi
Introduction	1
1. The Desert Biome of the World	8
2. Spain: Bardenas Reales – Iberian Badlands	39
3. The Netherlands: Dry Wilderness in a Watery Country	64
4. France: The Desert That Disappeared	91
5. Pustynia Błędowska: The Polish Sahara	112
6. Iceland: Cold Black Deserts	132
7. United Kingdom: Breckland – An Arid Desert in Verdant East Anglia	156
8. Germany: A Sandy Hike Through Purple Heather	190
Conclusion: European Deserts: Perspectives, and a Vision of the Future	215
Bibliography	229
Index	246

ACKNOWLEDGEMENTS

My sincere appreciation goes to the following for their generous and sustained support, advice, insights and critical comments offered during the preparation of the manuscript: Mark Linsley, Jeremy Stone, Caroline FitzGerald, James Pardoe, Andrew Marchant, Irene Stone, and Phoebe Morgan.

I want to extend my thanks to the innumerable conservation professionals, international academics, librarians, local authority officers and others who patiently responded to my requests for information.

I am grateful too to my supportive team of professionals at Pelagic Publishing, most notably publisher Nigel Massen for his support and encouragement. Also, production editor David Hawkins, marketing coordinator Sarah Stott, administrator William Reynolds, and copyeditor Sara Magness.

This book has been inspired by the work and achievements of the scholars who paved the way in this field and whose groundbreaking and inspiring work laid the foundation for this exploration.

All opinions, omissions, and errors remain my own and are not necessarily shared by any of the people or organisations that I have acknowledged.

INTRODUCTION

Europe is a green subcontinent. With a substantial proportion of the world's best soils under a generally favourable climate, agricultural expansion and intensification have resulted in the green and productive landscapes so familiar across much of the subcontinent. The EU is one of the world's major exporters of agricultural products, but food security has been achieved at the expense of traditional farming and the environment. As European land-use intensity has increased over the millennia, much of the natural environment has been modified – tamed, even – beyond all recognition, primarily for agricultural purposes but also for urban development and communications infrastructure (which now cover a full 4% of the EU's land area). At the global scale, too, modernity has brought ever-increasing expanses of the terrestrial biosphere into productive agricultural and other uses in many favoured locations.

Yet it is easy to fail to appreciate that Europe is atypical. Outside this green subcontinent, much of the terrestrial planet is barely or not at all productive. Less than half of the Earth's surface is significantly agriculturally productive, and it is instructive to appreciate that at least 40% of the global land area is desert. There are thousands of deserts on Planet Earth, distinctive landscapes where relatively extreme conditions prevail and little or nothing grows. On a global scale, a handful of particularly disadvantaged countries are classified as wholly desert: permanently arid and dominated by moisture deficit and low soil water availability, these parched places include some of the world's least-wealthy nations. To the casual observer in verdant Europe, the world's deserts may seem comfortably remote, but such a perception is an uninformed one. Firstly, the Sahara – the biggest and best-known hot desert – is disconcertingly close; so mighty are its powers that it influences the climate of the nearby subcontinent. But, secondly, the reality is that deserts are distributed all the way from the equator to the poles.

Europe has deserts. This book investigates this seemingly incongruous phenomenon for the first time, establishing and elaborating upon the notion of there being deserts in Europe, a green subcontinent. The work is founded upon the premise that the deserts of Europe may be distinguished with reference to soil aridity – also referred to as edaphic aridity – rather than high ambient temperature, the environmental variable more conventionally employed. A scarcity of available water is a more fundamental variable identifying the planet's desertic tracts, and soil aridity manifests in cold as well as hot climates. Deserts are relatively scarce in Europe but, nonetheless, they exist; not everywhere is favoured with productive soil and lush landscapes. As we shall see, instances of locations with soils with only limited ability to retain moisture are scattered across terrestrial Europe, varying in extent from relatively small to entire regions. The rationale for the book's use of the term desert for tracts of relatively arid substrate is more comprehensively discussed in Chapter One.

Europe's deserts offer a rare glimpse of the planet in the raw. These barren tracts are distinguished from their more familiar and productive surroundings by the dominance of the natural environment and natural processes. These 'cold deserts' offer a rare glimpse of unfamiliar and unruly wilderness on the green subcontinent. Frequently remote, unkempt and seemingly abandoned, the European deserts are comparable in many ways to the arid regions of the planet like the Sahara. Images of the Sahara are typically characterised by vast expanses of sand dunes, but while a few of Europe's deserts do feature dune landscapes the reality is that most deserts are not dominated by shifting sands: even the Sahara comprises only about 15% dune fields. The world's deserts are more typically grubby, monotonous plains of sand and gravel with tracts of dusty or bare rock pavement (but also ice, as we shall discover). These deserts are Europe's representatives within the global desert phenomenon, sharing the general characteristics of those elsewhere including their location on some of the worst soils – shallow, poorly developed, or wholly absent. Mostly free-draining and acidic, they are low in nutrient status and hence most are equally unproductive and unpopulated. Modernity has had little impact in these desolate places, and vistas and atmospheres in the European deserts are starkly different from their intensively managed surroundings.

These desert-like territories are the Cinderellas of European semi-natural habitats. Desert has been a feature of the subcontinent's terrestrial environment for millennia, and most or all were much more extensive in former times; what we see today are the remnants of formerly much larger tracts. Yet they remain mostly unknown and unrecognised, being routinely overlooked, neglected, unvalued and disdained. While many existed before humanity's influence became significant, most are at least partially cultural landscapes, having been produced and maintained over thousands of years by humans and their livestock alongside natural processes, hence the term 'semi-natural'. They remain an integral component of Europe's range of terrestrial landscapes and environments and comprise a collective heritage survival, yet our knowledge of their origins and the historic roles they fulfilled in local and national economies and societies is often remarkably scant. For the inquisitive and imaginative their status as novel and distinctive environments worthy of investigation is confirmed both by their stark otherness in relative terms and indeed their very obscurity. They should be regarded individually and in aggregate as a distinct and discrete phenomenon. Their rough, scruffy, yet often magical landscapes are as worthy of stewardship as any of Europe's better-known and more celebrated countryside regions, yet only rarely are they lauded in art, literature and national histories compared to other, more conventionally beautiful environments. Many also have reputations as undesirable, risky and even dangerous locations, and some of the people living in the European deserts have long been regarded in a similarly negative light, too.

Deserts are important for their contribution to the terrestrial environment and European and global biodiversity and geodiversity. In the twenty-first century, most of them are essentially abandoned and redundant in economic terms, unlike most of the subcontinent's intensively farmed land surface. With a scarcity of the chemical elements essential for a varied vegetation of higher species to develop, they are unable to support rapid growth, high biomass densities, or high biological diversity. Instead, they provide refuge for a limited variety and low numbers of specialist rather than generalist species of flora and fauna. Markedly dissimilar to that prevailing on the lands surrounding them, most European desert wildlife has nowhere else to go in the productivist countryside of the twenty-first century, where state-supported intensive agricultural policies with the goals of increased productivity and output take precedence over conservation. Characteristic faunal species include birds, reptiles and insects, many of which are dependent upon the presence of sandy open ground to hunt or lay their eggs, and flowering plants include those few adapted to such depauperate environments. Geodiversity is the abiotic (non-living) equivalent of biodiversity, contributing to the natural diversity of Planet Earth. Europe's deserts are notable for the distinctive landforms, sediments and soils which serve as the physical setting for life there, and merit conservation for their Earth heritage natural capital as much as for their biodiversity. However, while some are legally protected as nature reserves, overall many European deserts do not attract the attention of the various national conservation organisations.

The European deserts are worth cherishing for their human as much as their natural history. These are largely cultural landscapes, fascinating remnants of the subcontinent's shared human history and heritage. The deserts are as worthy of preservation and celebration as much grander human-made cultural artefacts like Europe's numerous palaces and cathedrals, and for many of the same reasons. These places have fascinating stories to tell about the people who once lived in these seemingly inhospitable rural backwaters. This book relates historical tales of extraordinary human endeavour, telling the stories of the skilled and hardy rural people who managed to win a living from these landscapes, sustained them into the twenty-first century, maintained and sometimes even created them:

- the Dutch *heide*-dwellers who for hundreds of years stripped countless tons of soil from the wastes to fertilise their fields (Chapter Three)
- the peasants who bounded around a vast French desert on stilts while knitting (Chapter Four); and
- the Polish miners who created a desert (Chapter Five).

This work aims to disrupt conventional perceptions, challenging prevailing assumptions about the European environment by contending for the first time that deserts are a reality on the green subcontinent. In a fresh interpretation, constituting the book's central contribution, these routinely unrecognised and unlauded arid territories are identified and acknowledged as deserts. While they are disparate in nature, exhibiting a degree of variation across the subcontinent, the deserts share sufficient common characteristics to be recognised as a collective phenomenon, one clearly distinguished from other biomes (bioclimatic zones, further discussed in Chapter 1), and a valuable one. This book advances the concept and term 'European deserts' which, in their conceptualisation, offer new perspectives on the terrestrial environment of the subcontinent.

Furthermore, it is contended that the strange has always been near at hand on the subcontinent: the deserts of Europe are comparable to the vast deserts strewn across more than 40% of the planet including the mighty Sahara. *European Deserts* introduces and paints a picture of elemental wildernesses, environments and landscapes that 'should not be there'. This book shows how in aggregate these places comprise a native component of the terrestrial environment, which generates novel insights for the jigsaw of knowledge on the European environment. They are empty, arid and even perilous but simultaneously fascinating, mysterious and somehow romantic, too. Their distinctive landscapes offer for European eyes a stark and dramatic contrast to the often carefully managed lands of their more familiar surroundings, and a tantalising flavour of the planet's vast arid deserts located on faraway continents.

So, what and where are the deserts of Europe? They are called, variously, heath, *heide*, wastes, *landes*, badlands and *desierto*, among many other terms. These lowland environments typically appear barren, with few significant visual features intruding into what are frequently open, sweeping vistas. Their colour palette is frequently anything but green, instead manifesting a dull, down-at-heel and even monotonous appearance for much of the year. They support only a very limited range of mostly unspectacular flora, and lack the conventional landscape beauty of many other types of European semi-natural habitats. Deserts are widespread in Europe; hundreds of such wilderness-like tracts, large and small, are scattered across the subcontinent, with instances located in almost every country. Yet 'desert' is a word only rarely employed in the literature on European habitats and landscapes. There has been a failure to acknowledge these dusty and infertile places individually as deserts, and a further failure to recognise the phenomenon of these deserts as collectively comprising a discrete desert biome across the subcontinent, a step requiring a more fundamental conceptual leap. In support of the book's thesis, the work is founded upon an understanding of desert applicable to the whole planet rather than just the hyper-arid subtropical desert regions of popular imagination.

Until now, Europe's deserts have in many ways been invisible in plain sight. It is a paradox that, while these deserts have been part of our landscape for millennia and, in their way, there is nothing like them on the subcontinent, few or none are widely known; yet many readers will be acquainted with one or more examples in their localities. While many of the largest deserts are located in remote rural areas, others are to be found on the fringes of and even within major cities, including London and Hamburg. Indeed, many European villages, towns and cities have tracts of scruffy and often unmanaged open space used for peaceful informal recreation and dog-walking in the fresh air. Yet few recreationalists would know how to refer to these places – except perhaps as 'wasteland' – and fewer still would perhaps recognise, understand or value them as the relict environmental survivals that they are. Prime examples include:

- Bardenas Reales, an unpopulated and little-known desert backwater in Spain, the size of the Isle of Wight (Chapter Two)
- Iceland, a desert island in the North Atlantic – though only rarely recognised as such – with the largest desert in Europe (Chapter Six)

- the Breckland region of Norfolk in eastern England, one of the most important British areas for wildlife with the UK's only inland sand dunes and species more typical of the Mediterranean and Russian steppe (Chapter Seven); and
- the extensive Lüneburg Heide set on sand in northern Germany (Chapter Eight).

What are the origins of these obscure backwaters, and how have they survived at all on this green and highly developed subcontinent? What is their significance as islands of European biodiversity? What has been the role of humankind in their histories, and how did humans once contrive to wrest a living from these mostly infertile wildernesses? What is their status as a pan-European habitat and environment type; what judgements may be made about their present-day quality and condition; and what is their future in an uncertain world?

The remaining deserts offer a scarce glimpse of a pre-modern and pre-industrial world. Indeed, local communities have often retained ancient collective legal rights over their nearby deserts, unlike elsewhere where rights over land more normally extend only to individuals. In the past, many peasants would have passed their entire lives knowing only these eerie, blasted realms. However, historical accounts of European desert-dwellers and their societies are scarce, perhaps because these places and those who lived in them were often treated with contempt by outsiders and not regarded as worthwhile subjects of study. In addition, many or most of the inhabitants would have been illiterate, without the ability to chronicle the passing of time.

But a glaring environmental tragedy is quietly manifesting itself in Europe's deserts. While many are protected as nature reserves, their continued existence remains under threat; many are in danger. Even in the notionally conservation-minded twenty-first century, this distinctive biome is declining in quality and overall extent. This environmental tragedy has been proceeding for hundreds of years through a complex combination of both natural and human-induced changes, a majority gradual and incremental but some large-scale and dramatic, though all mostly out of the public eye. Tellingly, another name and description applied to deserts all around the world is *wastes*, and accordingly those in Europe have long been regarded by many as worthless and even ungodly, an attitude which perhaps underlies a lack of effective restraint upon destructive influences in many locations. The most obvious damaging impacts are from the cessation of traditional management practices, agricultural encroachment, afforestation, sand gravel and mineral extraction, garbage dumping and wildfires. They are attractive locations for building houses or industrial or commercial buildings, offering sites which are level, stable and well drained. Some have been developed as golf courses, though with appropriate management elements of their wildlife interest may be retained, as is confirmed by some having been designated as nature reserves. Some are threatened by excessive visitor disturbance, especially those close to urban centres. Countless impacts have reduced the number and condition of these landscapes and resulted in habitat fragmentation.

General neglect has abandoned very many to the process of ecological succession, whereby one biotic community is progressively replaced by a new one. State-sponsored projects serving productivist agricultural policies over the last 200 years have led to large-scale 'reclamation' and loss of these areas in numerous European countries. Less readily perceptible but no less deleterious influences include atmospheric pollution and climate change, both of which have had significant effects. The impacts of these broader environmental changes seem likely to accelerate in the future. Founded upon the assertion that, over time, human activity has overtaken natural processes to become the dominant influence upon the planet, the Anthropocene thesis may be of relevance here. The genesis and continued existence of these predominantly cultural landscapes may offer valuable perspectives upon this contested thesis. These unique places have long been treated thoughtlessly: neglected, abused and abandoned in some countries, and in the inevitable competition for resources for environmental management more familiar and conventionally attractive habitat types have frequently been assigned a higher priority.

My inspiration and motivation for writing this book was based upon a lifelong interest in the environment. A native of north-east Essex, I grew up very happily in a lowland British village which had historically been part of a formerly extensive European desert. The soils are dominated by glacial sands and gravels, generally acidic and of low nutrient status, and surface water is a rarity. Discovering lizards and Grass Snakes among the sparse gorse and broom of some of the village's remaining fragments of heathland was exciting for an outdoorsy seven-year-old determined to understand the world around him. I came to appreciate the distinctiveness, beauty and rarity of my surroundings, but could only watch as the remaining semi-natural heathland sites were progressively destroyed. The Slow Worms and frogs disappeared under new houses, and thoughtless destructive behaviour and general neglect helped destroy the remainder.

I needed to understand why this biodiversity and landscape disaster on my doorstep was happening, and in such a seemingly unrestrained manner. When the time came, then, I elected to study environmental studies for my first degree, one of the earliest British environment bachelor programmes. After graduation, I was thrilled to gain an early career position with Suffolk Wildlife Trust. Part of an enthusiastic and well-qualified project team focused on the conservation of the European deserts of rural east Suffolk, ours was the first heathland project there, working to survey, assess and produce management plans for the string of sites comprising the so-called Sandlings. I was staggered at the sheer extent of east Suffolk's sandy heathlands, their characteristic flora and fauna, and intrigued at their collective survival. At last I was saving the world, and delighted to be working towards biodiversity conservation in however limited a fashion. Driven to understand these deserts, the realisation slowly dawned that the Sandlings heaths were just one collective instance of hundreds of such fascinating, slightly otherworldly, and largely unknown sites in the UK and indeed across Europe. Over the subsequent years I found myself unconsciously gravitating towards these places during holidays. Fired with the zeal of the enthusiast, I've visited some of the remotest arid backwaters of western Europe, on trips that invariably fed my interest and sense of wonder in the subject and helped build an appreciation of the scale and significance of the 'cold desert' phenomenon across the subcontinent. Over time, too, another realisation dawned: that I needed to write about it!

This is a mission-driven science, nature, environment, and society book based on professional experience and academic research with wide popular science appeal and a campaigning element. A modern-day environmental tragedy is proceeding in front of our very eyes. Most of Europe's deserts are endangered, and relatively few are cherished or actively conserved. Europe needs to act to ensure the sustainability of the subcontinent's remaining cold deserts with a comprehensive plan to conserve, manage and – yes – even expand these unique tracts, by means including rewilding projects such as those that have been pursued in the Netherlands and elsewhere. Presently the future for Europe's deserts is uncertain given the lamentably low levels of recognition of their existence and understanding of their significance. In the absence of much focus upon these unique environments and their wildlife, or a comprehensive strategy and plan with appropriately ambitious targets to assure their survival across the subcontinent, this book's final section presents the author's proposed vision for the European deserts in the year 2050, when:

- they are flourishing as an integral component of the subcontinents range of habitats;
- they exist at a favourable conservation status; and
- a target for the doubling of their collective extent has been achieved.

Chapter One

THE DESERT BIOME OF THE WORLD

Earth is the blue planet. Humanity inhabits its terrestrial areas, meaning that our species' focus centres upon the dry bits rather than the aquatic ones, but the extent to which the aquatic realm dominates the surface of the Earth is easy to overlook. Most of the Earth's surface comprises water, in aggregate an immense natural phenomenon whose extent, at over two-thirds of the area, dwarfs the combined total of the other four major planetary biomes: grassland, forest, tundra and desert. The Pacific Ocean alone covers an unimaginably huge one-third of Earth's surface area, accounts for about one-half of the world's oceanic water, and has a major influence on weather systems around the planet. Of course, the presence of available water is essential to the existence of life as we know it, and not just on Earth: when investigating other planets for the existence of life, scientists' search criteria invariably include the presence or absence of the water molecule. On a more everyday level, the curious, compelling attraction water wields over humanity is illustrated by the fact that about 40% of the world's population – nearly 2.4 billion people – live within 100 km of the coast. Water is a necessity for Earthly life, yet it is easy to overlook the fact that the liquid is in very short supply indeed over much of the planet's land surface, most notably of course in the planet's drylands and deserts.

Europe is a green subcontinent. The environment of Europe is characterised by verdant vegetation cover, largely driven by the temperate humid climatic conditions prevailing over much of the subcontinent, ranging from maritime temperate in the west to continental temperate in the east. The European climate benefits from the invisible but powerful Atlantic conveyor comprising the Gulf Stream, the most important ocean current system in the northern hemisphere, and its eastern extension the North Atlantic Drift, which together moderate the climate by transferring billions of joules of heat energy from the Caribbean subtropics to western Europe. Without the Atlantic conveyor, average London temperatures in December would be reduced by 5 °C, and overall, the subcontinent would be as cold as Canada at the same latitudes. Along with this favourable warm-water heat input comes large quantities of moisture-laden air, bringing water in the form of more or less reliable precipitation. The resultant humidity, in combination with large areas of favourable soils, once supported a natural vegetation largely comprising green, dense, species-rich woodland stretching from the eastern shores of the North Atlantic to the Ural Mountains, whose peaks mark the generally accepted eastern boundary of the subcontinent.

Favourable environmental conditions across western Europe have helped facilitate the historic development of productive agriculture. East of the Urals, environmental conditions are much less favourable: north-east lies the vast, frozen tundra of northern Eurasia; due east stretches the endless, sweeping steppe grasslands of central Eurasia, essentially semi-desert; and to the south-east are the margins of the south-west Eurasian Middle East hot arid zone. All are zones of relatively low biological productivity, with water constituting a key limiting variable in the latter two types.

Desert is the Cinderella of the world's environments. Obscure and neglected, these large areas of bare soil with a relative absence of vegetative cover exist at the opposite extreme from the blue and green environments elsewhere dominant over the planetary surface. Deserts everywhere are marginal, liminal backwaters: demographically, economically, geologically, geographically, culturally, politically, and socially. Farming in the most extreme infertile desertic locations is very limited or non-existent, rendering these places yet less in a world where production is paramount: most desert barely registers on conventional economic output scales. Human cultural landmarks are few and far between, too.

Desert landscapes are frequently regarded as featureless, boring and ugly; relentlessly unremarkable; a huge yawning nothing; and only rarely are they eulogised with the vocabulary of beauty applied to vistas of more familiar and conventionally appreciated environments like forests, mountains, rivers and lakes. In appearance, desert typically manifests a severely limited monotone spectrum of colour: grubby greys, dirty khaki browns and dull tawny colours far removed from the heady variegated blue and green palette prevalent elsewhere. True desert regions typically manifest a limited range of biodiversity and with relatively low population numbers, and few or no large or spectacular wildlife species in comparison to other major world biomes. Yet it would be remiss not to acknowledge the fact that desert is admired at least by some for its intrinsic qualities, notably including the singular starkness of the environment as epitomised by the Sonoran Desert of north-west Mexico and south-west United States of America, the Gobi Desert of Mongolia and China, and the almost totally uninhabited Namib coastal desert in southern Africa, the name of which derives from the Nama language meaning 'the place where there is nothing'.

Desert can be terrible: unknown, uncivilised, disdained, deprecated, dangerous, even despised. It can excite awe, terror and even dread. In some, the sun is pitiless, horizons are infinite, vistas almost limitless, the few physical features mutable day to day, and the emptiness seems boundless to people more accustomed to the small-scale landscapes and constrained horizons of the settled world. Some deserts are among the planet's last remaining areas of total wilderness. Desert is risky, and can be hostile: that dehydration, exhaustion and starvation are ever-present threats in hot deserts is made clear in the accounts of the British adventurers of the 1930s and 1940s who set out to cross the Arabian Peninsula's Empty Quarter or Rub' al Khali by camel. Bertram Thomas, Harry St John Philby, and Wilfred Thesiger each mounted their own camel-borne expeditions over the vast pile of sand that is the world's largest contiguous sand desert, sprawling over the United Arab Emirates, Saudi Arabia, Oman and Yemen, often against the best counsels of their local guides. St John Philby's journey involved a leg of 644 km between water sources. The diaries of these pioneering – if reckless – twentieth-century adventurers reinforce popular perceptions of the near-absolute lack of sources of water in this super-dry hell, with that available sometimes requiring excavation to depths of 60 m or more to reach. Even then it might not be reliably potable, instead frequently bitter or salty and sometimes polluted by the excreta of hundreds of years of camels assembling at the surface. Surface water here was even less common, but oases were often shunned by local populations because of malarial mosquitoes. And, in a novel paradox, even a surfeit of water in the desert can bring lethal danger, such as when a *wadi* – the traditional Arabic term for a normally parched desert valley – floods after heavy rain, bringing the threat of a sudden and violent inundation for the unwary. Death by drowning: in the middle of a subtropical desert! Desert may be impassable, regular movement constrained by highways which are indistinct, in poor condition, or more often absent, treacherous off-road surfaces lacking traction, and with food and fuel stops few and far between. Sandy soils are widespread across the world, but despite its nature as a common, humble and seemingly innocuous granular material, sand may pose a genuine hazard to human existence, capable as it is of moving in a fluid-like manner under windy conditions, ready and able to engulf farmland, rivers, lakes, roads and buildings in a natural process which may visit disaster upon *Homo sapiens* and their settlements and infrastructure.

Desert is mostly unpopulated. Globally, population density decreases as aridity increases, and United Nations estimates indicate that only a tiny fraction of the world's population calls deserts home. Population densities vary from 10 to 20 people per square kilometre, impossibly sparse compared say to Europe's 112/km^2, and settlements are scarce. Levels of human wellbeing tend to be lowest here: most people are poor – many subsist on less than US$1 per day – infant mortality rates are high, investment is low, gross national product (GNP) per capita is the lowest, and the United Nations considers that the livelihoods of more than one billion of the poorest and most marginalised people across some 100 countries are threatened by desertification, the process by which the biological productivity of arid and semiarid lands is reduced by natural and/or human causes. However, these regions are not often viewed as a priority for development either by dryland nation governments or international agencies, and in more general terms the global drylands and their uses receive inadequate recognition and attention, and their management suffers as a consequence.

Desert often lacks clear political or administrative borders, and often the obvious means of enforcing law and order in the wider sense. Many deserts extend across international boundaries, and many exhibit boundaries which are geometric and not always closely related to physical or cultural differences. Civilised authority in the form of national police forces frequently has its focus elsewhere, and deserts may lack effective boundaries and conventional means of law enforcement. In these ways, most or all deserts contrast dramatically with the everyday experience of outsiders, being perceived as risky compared to elsewhere, perhaps even existing outside the conventional realm of law and order where social conventions may not prevail or be enforceable. This is something well known to the British of the 1930s and 1940s whose camel-borne Arabian Peninsula expeditions suffered frequent attacks by marauding bands of armed bandits;

these desert adventurers were prepared for such attacks and seemed to accept them as mere inconveniences. From the relative safety and comfort of their day-to-day environments and societies, however, it is understandable that many people might be nervous about these wild, remote and relatively unregulated regions, where a state of anarchy may seem to be the norm.

DESERTS AND DRY LANDS

While the Atacama Desert is the driest non-polar desert in the world, the Sahara is perhaps the best-known example of hyper-arid conditions, the so-called true deserts of the popular imagination. The size of the continental USA and sprawling across thirteen North African countries, the Sahara is the world's largest hot (subtropical) desert by far, where persistent large high-pressure masses of dry air prevent the penetration of rain-bearing storm systems. Its very name is derived from the Arabic noun ṣaḥrā, meaning 'desert'; ṣaḥārā' is the plural. The absence of moisture in the air means few clouds in the sky and consequently higher solar irradiation. Annual sunshine can reach more than 4,000 hours in the Sahara; compare that to London with about 1,600. It is one of the harshest environments on Earth, where summer temperatures once reputedly attained 51.3 °C at Ouargla, Algeria in July 2018 – both the Sahara's and Africa's hottest reliably measured temperature. Rainfall figures lack much meaning over an area spanning roughly 4,800 km west–east by 1,800 km north–south, but water is scarce across the entire region and its extreme aridity is confirmed by annual precipitation averages varying from 75 to 120 mm. Compare these figures with those for London with 600 mm, and sunny Athens with 380 mm. Rainfall is highly variable everywhere in the Sahara, with none falling for consecutive years over large areas; much precipitation fails to reach ground level, evaporating as it descends from cooler through hot layers of the atmosphere; and at high elevations some precipitation falls as snow! The scarcity of water restricts the diversity and amount of plant cover to little more than a few scattered shrubs, and in turn the diversity and abundance of animals. The French term *désert absolu* refers to the most extreme sandy deserts, huge plains of yellow sand where nothing grows. But nowhere in the Sahara is rain-free, even in the half of the desert classified as hyper-arid; and most of the desert areas of the world are characterised by a wet season: that is, there is a predictable season of rainfall but the amount of rainfall in that season is not predictable. In many of even the driest regions of the Sahara, water may be present in its gaseous phase.

There's much more to this formidable hot desert than simply vast stretches of uninterrupted sand. While *ergs*, sand seas covered with dunes up to 180 m in height, are the dominant image of popular imagination, dune fields account for a relatively minor proportion of the Sahara's total area, which encompasses several ecologically distinct regions and a range of topographical features including extensive rocky plateaus; mountains rising to 3,000 m; plains of rock, sand and gravel; salt flats; basins; depressions; the remnants of former volcanoes and lava flows; and seasonal rivers. Fewer than 2.5 million people live here. The majority are nomads, predominantly the Tuareg, Tibbu and Moors, who survive by nomadic pastoralism, hunting and trading. There is no crop growth unless under irrigation. Most of the population is located around the margins of the desert, and the few green valleys such as the Nile; people spend little or no time in the central hyper-arid zone. The British government advises against travel to almost all the nations of the Sahara. The Sahara is not in fact the hottest place on Earth: the Lut Desert in eastern Iran, where in 2018 a temperature of 80.8 °C was recorded, holds that accolade. Remarkably, precisely the same temperature was recorded in 2019 in the Sonoran Desert, on the Mexico–US border. However, with a consistently hot footprint over a large area, the Lut has been declared the hottest place on Earth.

Inaccessible, inhospitable and risky, such extreme desertic environments may seem remote from the vantage point of the green and pleasant lands of western Europe. But their reality and some of their associated impacts are much closer than might be imagined. The northernmost bounds of the Sahara lie within less than 500 km of the more temperate subcontinent, and the desert's capacity for impacting European weather is illustrated by regular episodes of 'blood rain' in Europe: huge dust storms in the Sahara are swept north by winds in the upper part of the atmosphere to bring iron-rich particles as far north as Sweden, often seen as a thin film of rust-brown dust on cars and accompanied by distinctly un-European subtropical temperatures and humidity. Indeed, this uber-desert is the most significant global source of dust, accounting for half or more of all global windblown mineral dust emissions. An estimated 27 to 200 million tons is exported annually across the ocean from the Sahara to the Americas by transatlantic winds, causing air pollution problems in Caribbean islands and the southern USA. Some of this aeolian – wind-borne or wind-deposited – material helps to fertilise the soils of the Amazon basin, where the world's largest tropical rainforest benefits from the input of North African-derived minerals and nutrients – most notably phosphorous – which on deposition replenish soils depleted due to the intense competition for nutrients there. At a global scale, approximately 20% of the world's arid zones are covered by aeolian sand. Similar processes have been operating in western Europe over millennia – though on a much smaller scale – as wind-borne dust particles were swept from European deserts to engulf surrounding arable lands, often over durations of months or years. This stream of aeolian dust frequently rendered them unusable agriculturally, and caused settlement retreat on a temporary or permanent basis, a phenomenon which has had to be addressed in numerous European countries, as we shall see.

At this point, it is necessary to start differentiating the dry lands of the planet. The introductory discussion thus far has focused upon the so-called true deserts of the world, environments similar to the Sahara where hyper-arid conditions prevail. But as little as 6.6% of the land surface is classified as 'true' desert, indicating that such intensely hot environments are scarce. A more nuanced understanding stems from a recognition of two key points:

- that there exists a continuum of desertic environments, with the hyper-arid 'true deserts' of the world positioned at one extreme;
- much of the planet's terrestrial surface is recognisably desert-like, while not exhibiting the extremes of the Sahara or the Atacama.

Satellite image of the Sahara. (NASA, Public Domain via Wikimedia Commons)

Dry land is ubiquitous over the face of the planet, distributed not only across tropical latitudes but also temperate latitudes. The term drylands is used collectively for the roasting 'true desert' as well as the merely hot arid regions of the planet, regions where average rainfall is less than the potential moisture loss through evaporation and transpiration. Here, water scarcity distinguishes these areas from the planet's humid (or mesic) terrestrial zones at the opposite end of the moisture scale, the 34% of the total land area where moisture loss is equalled or exceeded by rainfall. The sheer scale of the global drylands is easy to overlook. Extending over a full two-fifths of the terrestrial globe, dryland constitutes the second-largest terrestrial biome – after the taiga or boreal forest of the high northern latitudes set between tundra and temperate forest – yet one about which we hear and know very little. The United Nations classifies the hot drylands into four climatic zone classes ordered by decreasing aridity:

- hyper-arid – the Sahara and similar
- arid – 'semi-desert' regions frequently bordering hyper-arid regions, including the semi-desert drylands of much of central Australia
- semi-arid – also termed pampas and prairie, these broad regions of typically grassy lands including the US Great Plains and parts of Spain are too dry to support trees away from water sources
- dry subhumid – frequently referred to as rangelands and similar to the Pampas of Argentina and the northern Eurasian steppe.

Hyper-arid zones exist at one extreme of the drylands continuum, where, with an aridity index of 0.05 or drier, potential evapotranspiration rates (the water that would evaporate if it were available) exceed the amount that falls by a factor of twenty or more. The exceptionally high evapotranspiration regime producing this massive moisture deficit in hyper-arid zones is powered by incoming solar radiation of such intensity that in hot-desert Qatar the energy delivered to each square metre of land surface is more than double that in northern Europe.

Represented in both hemispheres, on every continent, and in about 100 nations, drylands everywhere are defined by a scarcity of available water and characterised by low and unpredictable rainfall and cyclical droughts, high temperatures, high evapotranspiration, low humidity and seasonal climatic extremes. Many large dryland deserts are located in inland areas far away from coasts, where moisture from the oceans rarely penetrates. They are most prevalent in North Africa and Eurasia, though the western half of the USA is dryland, as is most of Australia and most of southern Africa and South America. Distributed around the mid-latitudes of each hemisphere, perhaps the most dramatic example is the vast belt of drylands sprawling from western Africa all the way to East Asia in a great yellow and brown slash across the face of the planet. In total, 63% of all arid and hyper-arid desert lands are located in the Palaearctic, the largest of the Earth's eight biogeographic realms comprising North Africa, Europe and the entire Eurasian landmass north of the foothills of the Himalayas all the way east to the Bering Strait separating Russia and the USA. The global desert-dryland biome is an intensely challenging one for plants and animals to survive and thrive in: vegetation is sparse, trees are usually absent, and shrubs and herbaceous plants may provide only very incomplete ground cover.

Globally, six countries are considered essentially 100% dryland in the sense of being arid or semi-arid, where the climate is characterised by a more or less permanent shortage of available water: Botswana, Burkina Faso, Iraq, Kazakhstan, the Republic of Moldova and Turkmenistan. Viewed from the humid world, it is easy to fail to appreciate just how far water availability controls the existence, distribution, and density of flora and fauna, including humans. Water availability is a major constraint on land use, dramatically limiting the available cultivable area and the production of crops, forage, wood and other goods produced or provided by biological productivity. Soils are often of low fertility status due to low water retention capacity, low organic matter content, and low nutrient content – particularly nitrogen – and are vulnerable to wind and water erosion, all making for low plant productivity which declines further as aridity intensifies. However, it is important to recognise that many of these bone-dry lands are cool, not hot: while arid, such regions are termed cold deserts, located as they are well north of the global subtropic climatic zones. Places like the central Eurasian Karakum Desert east of the Caspian Sea in Turkmenistan, a vast, sparsely populated territory of sand dunes one-third larger than the UK, experience long and cold winters with snowfall, and summers that are short, moist and only moderately warm, yet are still dominated by moisture deficit and low soil water availability.

But while it is easy to imagine that drylands everywhere are necessarily barren and of little value or interest, it is also mistaken. Once again, and like the point made about the Sahara, while many of these landscapes exhibit endless expanses of rolling sand dunes, gravel and rock, the world's vast drylands are markedly heterogeneous. A wide diversity of habitats and landscapes are distributed throughout the biome at the global scale, including forest and woodland, lakes, rivers, savanna, steppe, mountains, basins, salt flats, scrublands, shrublands, grasslands and semi-deserts as well as cultivated regions. All provide home range for a variety of specialist plants and animals adapted to tolerate the stresses imposed by unfavourable conditions including inconsistent rainfall patterns and, in many cases, high temperatures. Biodiversity may be limited in the hyper-arid regions, but is much more diverse in the arid, semi-arid and dry subhumid regions, many of which support an array of unique and well-adapted wildlife. One-third or more of the world's cultivated plants, including wheat, and many livestock breeds – horses, sheep, goats and cows – originated in drylands, meaning that these territories constitute an important genetic reservoir for the future.

With a population of 2.1 billion, the global drylands are home to more than one in four of the world's people. Three-quarters of the world's drylands are predominantly located in the less- and least-less economically developed countries of the world and, once again, dryland populations lag far behind the rest of the world on critical human wellbeing and development indicators. More than half of the world's productive land is dryland, with the predominant land uses being pastoralism and small-scale food production. Pastoralism is well adapted to the challenges of coping with scarce natural resources and environmental constraints such as drought. Food production is mainly from smallholding rain-fed systems for subsistence or local markets. Many pastoralists and sedentary farmers work together by exchanging crops and meat. But a rainfall regime that is low and variable still constitutes a high risk for agriculture and animal husbandry.

Once again, the global drylands are much closer than might be imagined. While much of Europe enjoys living conditions that are verdant and abundantly provisioned with water, no less than one-quarter of the subcontinent's land surface is classified as dryland under UN criteria. The conventionally defined European drylands are situated mostly around the Mediterranean region, but several tracts are located in central, eastern and even northern Europe, including a handful of isolated areas of dry subhumid climate situated as far north as Poland. The European deserts are located about halfway between the planet's cold and hot desert zones in latitudinal terms.

Aridity

The next major element necessary for fully establishing the concept of the desert derives from an exploration of the full meaning of the term arid. Here we must address a common misapprehension about deserts. Meteorologically, the term aridity centres upon water availability. Aridity is a climatic phenomenon characterised by a pronounced lack of available moisture. Meaning the temporal and/or spatial scarceness of water, aridity is a common characteristic of deserts, and the term arid is conventionally – and correctly – applied to the world's hot deserts. Regions may be described as arid if they manifest a relative scarcity or lack of moisture. Land in arid regions may be unproductive or barren because a lack of available moisture hinders or prevents the growth and development of animal and plant life, including woody plants and trees. But it is important to acknowledge that aridity is also a feature of the world's cold climate regions too, however counterintuitive this may seem. Aridity is characteristic of deserts located in both hot and cold regions of the planet, all the way from the subtropical desert latitudes to the poles. A United Nations definition identifies deserts as areas where vegetation is scarce or absent because of:

- deficient rainfall, or
- edaphic aridity – soil properties influencing plant growth.

So, deserts may be identified with reference to either or both of two critical variables related to biological systems, both of which relate to water stress – the scarceness of water. Examining the UN definition, the first of the two variables is that perhaps most typically identified with the classic deserts of popular imagination: a simple deficiency of rainfall. However, the second – edaphic aridity – needs a little more exploration. The adjective 'edaphic' relates to conditions in the soil, especially as they relate to biological systems. So, the second element of this UN definition enables the identification of deserts with reference to *soil aridity* rather than simply a scarcity of precipitation. Under this formulation, an area with an arid soil – one experiencing the temporal and/or spatial scarceness of water – may be defined as desert. Significant and sustained soil aridity constitutes a sufficient condition to identify an area as desert. Soils in the world's hot deserts are of course frequently arid; but, equally, soils may also be dry, waterless and parched in cold climate regions throughout the world, including western Europe. (The range of meanings of the term arid includes a figurative one, 'devoid of value' – alongside 'inhospitable', 'barren' and 'useless'.) Much of the area of the subcontinent is classified as 'warm temperate; humid; cool-summer' under the Köppen classification system, and in many European regions precipitation is certainly not deficient! But within this picture of temperate moistness, edaphic aridity is a commonplace attribute of many locations across this green continent. Soil water scarcity may be caused by a deficiency of rainfall in some locations, most notably in southern Europe, but across the subcontinent it is also a result of the inability of soils to retain moisture and deliver it to vegetation. What happens to the water in the soil is a critical determinant.

Soil texture, too, is of critical importance as it affects both infiltration and the movement of wetting fronts. Coarse-textured soils rich in sand are characterised by high infiltration rates and rapid percolation, while more fine-textured soils higher in clay and silt tend to impede infiltration, which is why they are commonly considered to be superior for plant production. Soil- and landscape-level processes including wind and water in the drylands modify the availability of soil water and nutrients. Soil modifies the effectiveness of precipitation, water availability to plants, nutrient availability and the physical environment for plant establishment. The absence of water in arid soils means that little or no chemical weathering can take place. A low moisture regime means that much soil in arid areas is almost totally devoid of decaying organic matter and micro-organisms, and also that small soil particles are easily blown away, resulting in dry places full of sand and characterised by frequent water stress – causing wilting and disruption of normal plant growth processes due to their inability to access sufficient water – low organic matter content and low nutrient content, particularly nitrogen. Set on a subcontinent with a substantial proportion of the world's best agricultural soils, Europe's lowland deserts occupy some of the worst.

Note here also that high temperatures do not figure in the UN desert definition. High temperatures are an attribute of many deserts, of course, but heat stress alone is not a sufficient attribute to identify a desert. Heat stress – when plants are exposed to high temperatures, causing dehydration and affecting plant growth and development with potentially severe, and sometimes lethal, adverse effects – is secondarily correlated with desert ecosystems, along with limitations of food and nutrient resources, but water scarcity is the first and foremost discriminating variable. Aridity is a climatic phenomenon principally characterised by a shortage of water rather than being directly related to average temperatures, and therefore may – and does – occur in cold climates as much as it does hot. Millions of square kilometres of land around the world with poor or absent soils may be regarded as arid, including Antarctica, for instance. It may take a significant perceptual adjustment for the reader to accept, but much of our watery and green planet is actually very arid.

DISTRIBUTION OF DESERTS

At this point in the book, it is necessary to refashion our perceptions of the phenomenon of desert yet further, and in a way that might perhaps at first seem counterintuitive, preposterous even. The discussion thus far has been largely focused on the hot and dry desert regions of the world. But the deserts of the planet are distributed all the way from the equator to the Arctic and Antarctic regions at the northern and southern polar extremes of the planet, and the world's largest desert is located at the South Pole. Not somewhere in the subtropical zone, which includes North Africa and much of Australia. The greatest desert on Earth is not blazing hot but freezing cold: the icy wastes of Antarctica. The Sahara isn't the largest desert – Antarctica is. Larger than the United States, the South Pole is a vast desert with a polar ice cap climate and is extremely cold all year round, with dark, frigid winters and almost no vegetation. Set upon the permanent ice sheet 1.6 km in depth that dominates the continent, Antarctica is the coldest continent by far. Monthly average temperatures only rarely exceed 0 °C, and the coldest temperature ever recorded on Earth of –110.9 °C (colder than Mars) was recorded here by satellite in 2016. Just to add to the catalogue of continental extremes at the planet's southern extremity, Antarctica is also:

- the driest
- the brightest
- the windiest
- the highest in terms of average elevation
- unknown – the land surface underneath the east Antarctic ice sheet is less well known than the surface of Mars.

Antarctica is the driest continent on Earth. So dry is the Antarctic Desert that the continent is classified in the hyper-arid category alongside the Sahara and Atacama, despite the seeming paradox that 60–80% of the world's fresh water is locked up in the ice at the South Pole. A super-cold desert covers almost all of the continent, but in place of sand there is only compressed ice and snow, to a depth of up to 4.7 km. In a brutally low-temperature environment, the permanently dry air here yields a miserly annual average of 50 mm or less of water-equivalent precipitation over the interior, most of which falls as snow and ice crystals. There are also large snow and ice dunes: lying

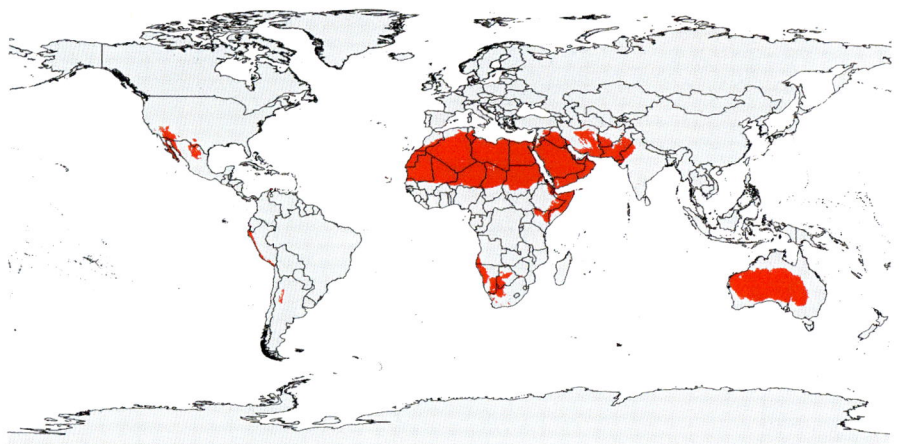

Hot desert climates of the world. (Beck, H.E., Zimmermann, N.E., McVicar, T.R., Vergopolan, N., Berg, A., & Wood, E.F., CC BY 4.0)

on the East Antarctic Plateau are extensive fields of 'megadunes', recrystallised snow in the form of long, wave-like ridges a few metres high. Formed by centuries of nearly continuous winds, from space their corduroy-like patterns look like giant fingerprints, and they drift across the ice sheet surface at a velocity of about 5 m per year.

So, while it may seem thoroughly counterintuitive, the Antarctic is as arid as the Sahara. How may this be so? We know that the Antarctic is a water-based world, a polar ice cap with an ice sheet covering about 98% of the continent. But this polar desert receives very little annual precipitation, and almost all falls in the form of snow and ice. Aridity here is because the limited icy precipitation fails to melt over most of the continent, instead becoming compressed and accumulating over hundreds and thousands of years thus forming the glacier ice that comprises the ice sheet. The key factor determining plant life in Antarctica is the availability of water. While we can recognise and understand it in its liquid phase, water in the Antarctic exists in the solid or mineral phase. Ice is the solid phase of water, commonly forming a structure of hard, amalgamated ice crystals or more loosely accumulated granular crystals, such as snow. In the temperature range critical for biological life (0–100 °C), water occurs in its liquid phase, but when in its solid phase at the permanently low temperatures prevailing in the Antarctic the molecules are bound together in a tight, rigid framework: a crystal. In the Antarctic, and many similar cold parts of the planet, not only is there rainfall deficiency but also edaphic aridity linked to this extraordinary water-as-a-solid phenomenon, and as a result the continent's abundance of H_2O is mostly unavailable to plants. Ice is not soil and does not support plant life. Only about 0.4% of the continent is permanently ice-free, where the climatic conditions support only relatively impoverished tundra vegetation of cold-tolerant land plants existing under conditions of low nutrient status and low water bioavailability.

There are relatively few species in this depauperate ecosystem – mostly mosses, and lichens, fungi and algae – very little total biomass and net primary productivity, and only two species of flowering plant. Most of the flora is located on the margins, along the coast and on the islands which represent hotbeds of biodiversity on the continent, where temperatures are milder and there is sufficient precipitation to enable plant germination and growth, and the fauna is essentially sea-based. This desolate and desertic environment is characterised by large variations in temperature and light regimes, meaning plants need to be capable of surviving extended winter periods with no sunlight during which photosynthesis is stalled, and when the sun returns it delivers a high incidence of solar radiation with an elevated ultraviolet B (UV-B) radiation component, linked to reduced plant growth. Not all deserts are hot; and aridity is a relative concept.

Yet more curiously, the South Pole also accommodates sandy landscapes indistinguishable from some of the planet's subtropical deserts, with some areas supporting few or no living organisms. Set within the east Antarctic ice sheet, the largest on the planet, are pockets of cold, dry and arid desert. Implausibly snow- and ice-free, the McMurdo Dry Valleys feature huge aeolian sand dunes and desolate rock-strewn dry valleys. Built by the wind, at 70 m in height and more than 200 m in width Antarctica's largest dune

The continent of Antarctica, c.2006. (Davepape, Public Domain via Wikimedia Commons)

is located in Victoria Valley. Gale-force winds driving over the eastern Antarctic and descending from the ice sheet can reach speeds of 320 kph and increases in temperature as it does so, creating bitingly dry arid conditions and evaporating all water, ice and snow in some locations. Some of the coldest and driest permafrost soils on Earth can be found in these Antarctic dry valleys, many of which have not had rain for one million years. In December 1903, when British Royal Navy officer Robert Falcon Scott and the Antarctic *Discovery* expedition first set foot on one, they described it as a 'valley of the dead' for its lifeless, barren quality. Conditions are extraordinarily hostile for life, and these Antarctic dry valleys are the first location on the planet visited by humans with no significant active microbial life. Some scientists consider the dry valleys perhaps the closest of any terrestrial environment to the planet Mars.

So, Antarctica is an uber-desert: a polar desert. The Antarctic ice sheet dominating the region is the largest single chunk of ice on Earth in a region with the most inhospitable environment on the planet, located far from all other continents, and unique in having no native population or human habitation other than scientists in research stations. Acknowledging that hyper-arid Antarctica and other cold and arid world regions are desertic arguably increases the cumulative figure for the global drylands. The conclusion to be drawn is that *nearly one-half of the planet's land surface is arid*.

The final premise necessary in fully establishing the concept of the European desert is a brief exploration of the nature, characteristics and significance of the planetary biomes.

Upper Wright Valley in the McMurdo Dry Valleys, Antarctica. (Pierre Roudier, CC BY 2.0)

Desert is one of the world's five major biomes. Biomes are *bioclimatic zones*, subdividing the planet into large-scale regional units of the natural world zoned with reference to common constellations of adaptations possessed by the organisms living there. Named after the type of vegetation dominating over each, the five basic major biomes are grassland, forest, aquatic, tundra and desert. Each is a system formed by an ecological community and its environment that functions as a unit. Their distribution around the planet indicates clearly that the same biome can occur in geographically distinct areas with similar climates, with for instance temperate forest prevailing in a far-flung distribution across Europe *and* south-west South America *and* eastern Australia. This concept of major life zones contributes to understanding the types of natural environments found on Earth and making sense of their complexity. Plants and animals living in each biome possess common and distinct constellations of adaptations to them, with climate perhaps the most important element in determining which kinds of organisms can live in a region. Other significant factors include geology, soil types and processes and light and water availability.

Dominated by trees, forest is a major global biome covering perhaps one-third of Earth's surface. Forests contain much of the world's terrestrial biodiversity, including insects, birds, and mammals. At a global scale, under a mostly temperate climate, the European subcontinent is dominated by the forest biome, although many of the trees were chopped down a long time ago. But investigate a little more closely and a more detailed picture of the subcontinent emerges. Biomes exhibit significant internal variation, encompassing a range of climatic and soil regimes and multiple ecosystems with their various associated communities of plants and animals. The majority of the lowland European subcontinent, including the UK, is conventionally zoned as a region of temperate broadleaf and mixed forest biome, with southern Europe around the margins of the Mediterranean basin zoned as 'Mediterranean forests, woodlands and scrub'. But the large scale at which biomes are demarcated affirms that the concept offers only an approximate representation of the situation on the ground. Only recently was it established that 1.8 billion trees are growing in some of the most parched areas of north Africa. The equivalent of another Amazon rainforest is flourishing in climatic zones which vary from semi-arid to hyper-arid across parts of the West African Sahara and the Sahel. This phenomenon of trees growing at relatively high densities in desert regions – where they have a crucial role in biodiversity – challenges preconceptions. To restate a point already made, the world's arid lands are not confined only to the subtropics but are distributed across the globe and all the way from the equator to the poles, for their characteristics lie within the range of internal variation exhibited within many biomes.

Europe has a substantial proportion of the world's best soils in terms of arable agriculture, but such a benevolent endowment serves to make the subcontinent's least agriculturally favoured locations seem yet more anomalous. Even in verdant Europe, up to 24% of the land surface is formally classified as dryland. Hence a clear case may be made for recognising the concept of *European* deserts, locations exhibiting recognisably desertic characteristics *relative* to their setting on this green subcontinent. These relatively arid lands indisputably constitute European instances of the wider global desert phenomenon, given their *relatively* desertic nature compared to their surrounding areas. There are several precedents for such a seemingly extravagant interpretation, including for instance the case of the rainforests of the British Isles. The term rainforest is more often associated with tropical rainforests including the Amazon, but *temperate* rainforest occurs in many moist oceanic regions around the world, including the rain-soaked west coast of the British Isles from Cornwall to Scotland. The hallmarks of this moist, species-rich woodland habitat are oak trees whose branches and trunks are festooned with epiphytic plants – plants that grow on other plants, most common in moist tropical areas – including mosses, lichens and ferns and a dense understory of shrubs and berries. The UK's temperate rainforest is important for biodiversity, supporting rare and threatened species, and many sites are ancient woodlands. Sites such as Wistman's Wood on Dartmoor constitute British representatives of the global rainforest in the same way as the deserts in Europe compare to those around the world. Moreover, with parallels that are grimly similar, little survives of this damp ecosystem that has slowly been cleared by humans over the centuries and today is threatened by pollution and invasive species. Accordingly, a fundamental premise upon which this work is based is a recognition of the existence of distinct, identifiable locations in Europe which are broadly equivalent to the classic drylands of the planet. The phenomenon of desert on the European subcontinent is worthy of recognition and investigation, offering a novel focus for the inquisitive.

Despite being positioned geographically in a temperate climatic zone, there are numerous locations and even whole regions in Europe which exhibit many of the characteristics of Earth's drylands. Highly distinctive, they are unlike anywhere else on the subcontinent, and may justifiably be considered analogous to the world's so-called true deserts. So: where are the deserts of Europe? To offer a swift example, Germany has extensive areas of dusty *heide* heathland which, while not subject to a hot, hyper-arid subtropical climate, are deserts *relative* to their geographical setting insofar as they share many of the characteristics of the drylands recounted earlier. Broadly similar instances of European deserts are located in most European countries. Such dusty tracts have often been named 'deserts', both informally by local people, and in the literature and occasionally on maps too: in the Netherlands; in south-west France; in Poland; in the UK. Why do they exist, what are their histories, and how have they survived in such a highly developed world region?

ENVIRONMENTAL VARIABLES

Which are the fundamental environmental variables identifying Earth's deserts? Clearly, aridity may be one – but only one of several, and it is not the most fundamental. There are numerous climatically hot regions around the globe, but most are not desert: indeed, many are more verdant than Europe, including for instance rainforest in tropical latitudes which exhibits gloriously abundant vegetation and species diversity. The fundamental variable identifying desert environments is soil, or the lack of it. Soil is in many ways the Earth's most valuable resource. Suitably developed soil is essential for plant establishment, growth and reproductive success. All around the world, arid desert soils are poor. In arid regions, a combination of a high-temperature climatic regime, meagre precipitation, surface geology and limited vegetation cover combine to produce soil that is poorly developed or even wholly absent. Soil-building in hot deserts can be a very slow process indeed, and in consequence the plant-growing substrate there is unable to sustain abundant vegetation with higher plants because of an inability to retain moisture. In addition, prevailing high desert temperatures boost rates of evapotranspiration, in turn limiting the depth to which water may penetrate the ground and restricting the depth to which soil may develop. Wind is often another significant edaphic factor too, further contributing to water losses from evaporation and plant transpiration as well as acting as a major agent of erosion, roughing-up and burying the already sparse vegetative ground cover and, in contributing to the impermanence of the soil, reducing the prospects of there being any consistent substrate upon which plants may flourish. In regions of more temperate climate, a roughly similar combination of poor surface geology with low substrate fertility, seedling development processes proceeding at a slow rate, the effects of wind, and relatively high rates of evapotranspiration can also result in soils that are poorly developed or absent, and as unable to retain soil water moisture or support abundant vegetation with higher plants as many in the subtropics. In both arid and temperate humid environments, too, human intervention is frequently a significant factor; of which more later. So, while drylands are ubiquitous across the face of the planet, *deserts* are yet more widespread. Everything is relative in the natural world; the concept of desert varies along a continuum; and every continent has arid tracts which approximate to desert when considered relative to the continent as a whole. The whole concept of the desert needs to be reframed.

EUROPEAN DESERTS

Europe has deserts. Cool deserts, largely unacknowledged deserts, but deserts nonetheless. These deserts have been an integral component of the face of the subcontinent for millennia, and were far more extensive in former times. A central theme of this work is that many of the characteristics of the world's drylands are replicated in these territories so uncharacteristic of Europe. This work will introduce the concept of European deserts, and go on to investigate several significant examples, each selected to illustrate the diversity of these desert-like places across this green subcontinent. The deserts of Europe are to be found around all four points of the compass, from UK in the north-west and Poland in the north-east to Spain in the south-west and Romania in the south-east, yet only rarely are they acknowledged individually or in aggregate as the phenomenon they are and for what they represent. Few or no European deserts have gained any significant profile internationally, and not too many much domestic profile either – although Lüneburg Heath in German Lower Saxony is perhaps the best known, though not primarily for environmental reasons (see Chapter Eight). Another in south-east Spain, the parched and depopulated rock-desert of Tabernas, perhaps offers the closest resemblance in Europe to the global drylands, although even this remarkable region has failed to gain much recognition despite its striking environment. Taken together across the subcontinent, these are some of Europe's most distinctive, stark, mysterious and fascinating wildernesses, rare examples of relatively extreme environmental conditions set in the lowlands of this largely green and fertile subcontinent. They resemble no other inland European terrestrial habitats. These cool deserts are as deserving of our recognition, attention and support as any other semi-natural environment type in Europe.

European deserts are typically located on mineral-dominated sand and rock geological substrates. Many are located on the North European Plain, a vast glacial outwash plain that runs west–east over across northern Europe from eastern England and the Low Countries over northern Germany and most of Poland before sprawling eastwards into Belarus, Ukraine, and beyond. Soils throughout are poorly developed and sometimes essentially absent, free-draining, acidic and low in the nutrients essential to develop and support a varied vegetation of higher plants. In terms of physical appearance, the deserts vary geomorphologically with many lightly sloping or essentially flat in profile, and some even plateau-like. Others are located on sand and gravel ridges originating from glacial action. Some of the more extraordinary examples include the European deserts of the inland Veluwe region of east-central Netherlands, where plains of open sand and windblown dunes – very uncommon in inland Europe – bear a strong resemblance to the hot deserts of northern Africa. At the other end of western Europe, and set upon a very different geological substrate of clay, limestone and sandstone, the extraordinary Bardenas Reales region of central north-eastern Spain (Chapter Two) resembles the US deserts of Colorado with dramatic, sparsely vegetated and sparsely populated landscapes of rocky tablelands known as mesas, elevated areas of land with a flat top and steep sides like nowhere else in Europe. Such rock and sand substrates reflect about twice as much incoming solar radiation as do grassland or woodland, heightening the desertic feel for the casual observer.

One of the key attributes common to all these places is their lack of moisture. Terrestrial environments in general are characterised by limited water availability, making dehydration a threat. The lack of available water is a critical biotic factor characterising the European deserts, exactly as it does the better-known global drylands. Our everyday experience is of a green, moist subcontinent set on complex soils, some of which are water-retentive or occasionally waterlogged, perhaps making the significance of water availability easy to disregard; but the European deserts are different, where soils are generally dry year-round, and frequently parched in hot summers. In locations with well-developed soils, when rain infiltrates the ground a proportion is retained in the upper layers as the water's downward progress is impeded by humic layers (the organic layers of soil made of decayed leaves and plants) and other factors, with the result that the soil remains relatively moist for a period. However, precipitation falling upon European deserts rapidly infiltrates the coarse, loose and usually dry upper layers, and the ground may remain moist for only a short duration as there is little to arrest the precious fluid's gravity-driven progress downwards and out of reach of most plant roots. Many inland European heather heaths are exposed to summer drought, where plant-available water in the root zone is exhausted. The prevailing climate may not be arid, but when it comes to water availability these areas can be like drylands anywhere. Being located mostly inland and lowland, they generally receive significantly less precipitation than coastal and mountainous locations. These areas are often located in the upper reaches of watersheds, and water tables tend to be far underground, particularly in ridge locations. Surface water sources are infrequent and often only seasonal, and standing water is equally uncommon. Areas with higher water availability may typically be based upon perched water tables or where drainage is poor, providing suitable conditions for marsh – where Bilberry, sedges, rushes and reeds may flourish – as well as ponds and shallow lakes, and often support scarce amphibia including newts and toads, but these are usually only small in overall extent. Surface evaporation may be high. To support livestock grazing the provision of drinking water is usually essential, and prospects for human habitation are equally dependent on water availability, in the same way as in drylands elsewhere in the world. Overall, these areas provide a haven for specialised flora and fauna adapted to a low water availability regime.

Ecologically, these areas are unique in their unspectacular yet distinctive way. The European deserts rank among the least common of the subcontinent's terrestrial habitats and ecosystems. Geology and soils form the basis for a semi-natural habitat type with a distinctive floral and faunal composition and high ecological value and interest. Like the global drylands, life in these areas is adapted to a lack of water and nutrients. In comparison to elsewhere, the mineral environment dominates to produce relatively extreme environmental conditions. The geological substrate which elsewhere would supply essential inputs for the nutrient cycle is fully weathered: sand, gravel and in some cases bare rock have few or no nutrients to offer. Nutrient-poor soils are unable to support rapid flora and fauna growth, high biomass densities, or biological diversity in general terms simply because of the dearth of the chemical elements necessary to support life. Plant growth is mostly dependent upon nutrients liberated by the decay of existing vegetation, which in turn is in short supply in a cycle of nutritional poverty. The lack of available soil moisture compounds the challenges and further serves to distinguish these semi-natural environments from more fertile locations.

In consequence, most European deserts support a range of vegetation which though specialised is lacking both variety and numbers of species, and with relatively few higher plants compared to other semi-natural habitats. From an ecological and physiognomic view, many are communities in which dwarf shrubs are dominant though frequently sparsely distributed, mostly dominated by heather-type species. Developing over nutrient-poor soils, this vegetation is tough, scrubby, often evergreen, and sclerophyllous – plants with leaves described as small, hard and thick, adapted to arid climates. Sclerophyllous leaves developed as an efficient adaptation to the physiological stress caused by limited water availability, and are low in nutrients, a common characteristic of plants in nutrient-impoverished and/or droughty environments. Where they survive, trees may include pine and birch species, oak and juniper in some locations; invasive and frequently alien species seeding from nearby human-made plantations may also be in the mix. The more arid southern European desert regions provide habitat for tamarisk, a fire-adapted desertic shrub or tree native to Eurasia and Africa (and introduced to North America), so well adapted that it is capable of tolerating both saline and alkaline conditions, and able to access deep water sources. Over the drier areas, mosses and lichens may become dominant: they can produce a distinctively grey, scruffy micro-landscape. In the very driest areas, there is frequently little or no vegetation at all, the ground surface comprising bare sand or sometimes rock. So, the everyday ecological challenges faced by the flora and fauna in these locations are similar if not identical to those found across the planet's drylands: while very little of Europe is arid, one of the most extraordinary characteristics of these places is that while located on this green continent they are comparable to many of the world's dryland environments.

Alongside the range of generalist species associated with the habitat, European deserts' faunal profile features numerous species which have nowhere else to go. Many are threatened, including such birds as Nightjar, Dartford Warbler, Stone-curlew and Great Bustard, a bird more usually associated with the drylands of temperate Central and East Asia and once a common sight in European agricultural areas. Reptiles include snakes, lizards, Slow Worm and a very diverse range of invertebrates including Black and Red Sand Wasp, beetles, spiders, solitary bees and ants. Rabbit was introduced as a source of food and fur, and numerous warrens were established in northern European deserts, often huge open enclosures where the creatures lived almost wild. Today, significant rabbit populations continue to intensively graze these areas, alongside other endemic mammal species including deer, and anthropogenic introductions in some locations including sheep, goat, cattle, pig, horse and even wolf. Various invasive species adversely affecting locally native biodiversity are a problem in some locations.

Desert landscapes in Europe offer a dramatic contrast to more familiar verdant landscapes across the subcontinent, and indeed to most other semi-natural habitats. Predominantly open in character, the landscapes of these semi-natural environments present as recognisably desertic in nature in comparison to their surroundings. Few or no other locations offer such an atmosphere of wilderness in lowland Europe. Typically, they present with open, sweeping landscapes on flat, gently undulating, or rolling topography. They offer long views and huge skyscapes, often enhanced through the proximity of adjoining woodland that accentuates the feeling of wilderness.

The appearance of southern European examples, meanwhile, approximates more closely to the hyper-arid wilderness areas of North Africa. For some observers, however, their appearance may be judged unremarkable, dull, monotonous, barren and featureless, predominantly treeless, and with mostly unspectacular flora. Over much of the year, their colour palette is far from green; a spring splash from a scattering of flowers and herbs gives way to high summer browns, khakis, buffs and dull greys as the heat and dryness burns up landscape and flora. A dramatic late-summer flowering of the dwarf-shrub heather and similar species produces an attractive purple floor show here and there, and gorse, which flowers throughout the year, presents a welcome bright yellow splash, but in general the European deserts lack the conventional landscape and floral beauty of many other European semi-natural landscapes and habitats. Partly perhaps because of their unspectacular brown-grey colouration, European deserts have been deprecated throughout the centuries, and their appearance is regularly described as 'blasted'. Nonetheless, taken together these distinctive cold deserts constitute a common European landscape heritage, with examples to be found in most or all countries across the subcontinent.

THE HUMAN DIMENSION

Many or most European deserts are semi-natural rather than wholly natural. Most or all of Europe's environment has been modified by humanity to a greater or lesser extent; hence the term semi-natural. These arid lowland wildernesses have been substantially human-influenced over millennia, and even in the present day most are human-maintained. Characterised as secondary habitats, their continued existence is dependent upon sustained intervention. In post-Ice Age times, early human populations recolonising the vast expanses of densely forested European wilderness sought out and settled relatively dry places, frequently selecting locations with sand and gravel as the predominant underlying geology including former glacial outwash plateaus and ridges. Here, hunter-gatherers exploited the lightly wooded alluvial soils for food, wood, firewood and probably leaf litter too, collected as fodder for livestock during winter. As the Neolithic era progressed with the emergence of settled agriculture, woodland clearance proceeded on a large scale, and many present-day European deserts were carved out of the wilderness. The relatively easily tilled soils yielded land for livestock grazing, cropping and – very simply – room to live. However, set upon fully weathered substrates, the poverty of these low-nutrient soils meant that their productivity declined rapidly, paralleling the impacts of similar present-day deforestation processes in other world regions.

Before too long, humans began to move away from these early habitation locations as their fertility diminished. They headed for more productive lands, even though these typically required considerably more effort for clearance, drainage, soil preparation and tilling. However, the notionally abandoned desertic areas continued to play a role in the rural economy, exploited by nearby settled human populations for their scant remaining fertility. Open landscapes were maintained through extensive low-intensity grazing and foraging for food, firewood and whatever else was available to supplement farm production. Many or most of these semi-natural landscapes were alternately recolonised and abandoned at various times over the past 10,000 years in response to changes in population pressures, local and national politics and the climate changes of the Holocene (the geological epoch that represents the last 11,700 years). Thus, their development was heavily influenced by anthropogenic factors: their present-day biodiversity-impoverished state was entrenched several millennia ago by the intervention of early humans. Nutrient-stripping practices, coupled with the influences of wildlife and natural processes, perpetuated their low-fertility status, and are a distinguishing feature of these tracts.

Much of present-day Europe is intensively farmed, typically involving regular ploughing and applications of the inorganic fertilisers and pesticides essential for the cultivation of conventional crops. Agrochemicals are inherently persistent in nature and may have adverse effects on soil health, biodiversity and the environment in more general terms. In stark contrast, many or most of these desertic tracts may never have been exposed to agricultural chemicals, and at least some may have never been put to the plough. Today the paleolandscapes of the European deserts represent precious remnants of past environments. Such relatively undisturbed soils are scarce in the European lowlands, and qualitatively different to those of many of their surrounding lands. Twenty-first century low-intensity grazing there is essentially a heritage land use. Long established

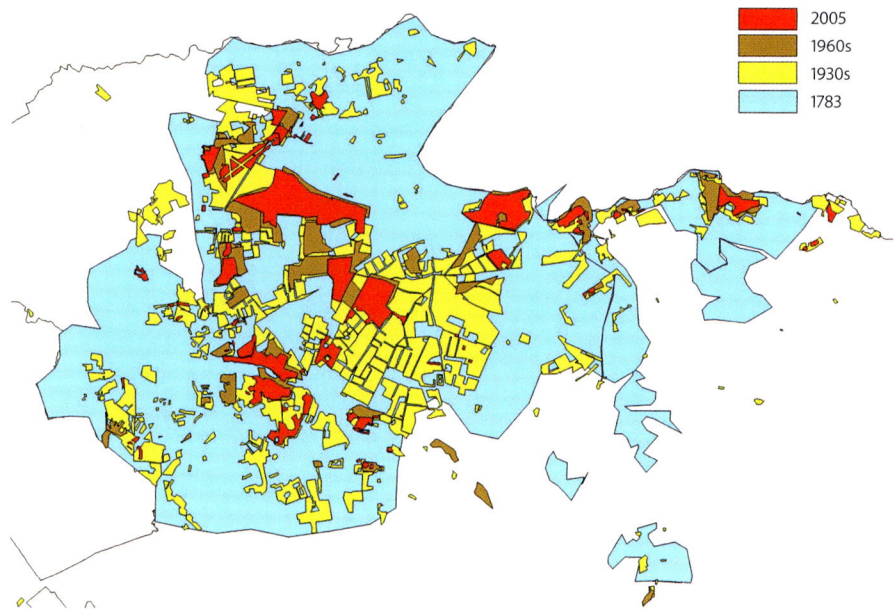

Decline in the extent of the European Deserts of Norfolk UK: The Breckland (see Chapter Seven). (Compiled by Martin Sanford)

management practices have served to maintain much of their biodiversity, and they fulfil a critical role as refugia for a specialised set of flora and fauna. They may also play a less appreciated but potentially significant role as local gene banks, reservoirs of genetic resources of species adapted to survival under arid environmental conditions.

A key characteristic of most European deserts is their economic redundancy; most serve few or no conventional economic purposes in the twenty-first century. Their poor soils are usually unsuitable for twenty-first-century agriculture, again contrasting sharply with the familiar, economically productive agricultural lands dominating much of their surrounding landscapes, and indeed much of the rest of the subcontinent. The best bits have been reclaimed for agriculture over the centuries, and in a situation of high agricultural output from more favourable land there is little incentive to attempt to cultivate the last few least-productive hectares. Some European deserts are used for low-intensity, extensive rangeland-style grazing, providing local farmers and individuals with some marginal income through sheep, goats, horses, or cattle; some are a focus for extractive industry, including sand and gravel quarrying; some are reserved for military use; and some suffer the indignity of being used as casual garbage dumps. Many also fulfil a critical role as nature conservation sites, whether by accident or design, managed or unmanaged, or simply by default on account of sheer neglect; and many close to settlements are popular as recreational sites, with a few including some of the Dutch and German heaths attracting significant numbers of tourists. But while land use in most of these areas is non-marketised in nature, this present-day picture contrasts dramatically with their former status; in later chapters, we will see how in previous centuries these seemingly economically redundant wastes were exploited on a systematic basis. In some places heathland was highly valued economically, maintaining a central role in the rural economies of parts of Europe for millennia until about a century ago. Today, and unlike most of the rest of the rural European real estate, their mostly redundant status in conventional economic terms is paralleled by the fact that many are abandoned in functional terms. However, that does not mean they have no owners: while it is inadvisable to generalise at a subcontinental scale, probably a majority are owned by one or more legal titleholders, and very many also have elements of legal rights of common attached – shared ownership and/or usage – including, for instance, the New Forest of southern England (much of which is open ground rather than being afforested as its name suggests), where commoning remains as a farming system and historic rights to shared grazing land and other land-based products are still exercised by some.

European deserts are frequently valuable archaeologically too. Compared to their surroundings, most have experienced less or even no human-generated ground disturbance or tree root intrusion of the sort that has damaged or destroyed innumerable ancient archaeological sites in the fertile European lowlands. Thus, many are very important for the number and range of ancient monuments preserved within their bounds, and the relatively dry environment is often favourable for the preservation of barrows, tumuli, hillforts and the remains of prehistoric economic activities, notably including the unrivalled Grimes Graves in Norfolk – not graves at all, but Neolithic flint mines. Remnant mire and bog habitats surviving in these areas often yield valuable preserved pollen profiles, while those located in other landscape types were swept away by 'improvement' a long time ago. They also represent the last vestiges of certain historic European landscapes, for unlike almost anywhere else one can experience a landscape on the subcontinent's deserts which early humans would have recognised. In his classic book *The Making of the English Landscape* (1955), English historian William George Hoskins wrote how exercising one's imagination on a 'desolate moor' can transport the individual back in time to the Bronze Age, a sentiment particularly relevant to the European deserts, many of which may not have changed markedly in appearance since then.

Bad places, bad people: in a weird phenomenon, sociocultural attitudes towards and perceptions of the European deserts and their populations have long been strikingly similar across the subcontinent. The term desert is derived from the Latin *desart* meaning abandoned, and while much of Europe is densely populated and predominantly urban, the deserts of Europe are invariably very rural, mostly thinly or unpopulated and frequently located in underpopulated and geographically remote regions with few settlements or highways. In another echo of the defining characteristics of many of the world's arid regions, desert environments in Europe have frequently been disdained, deprecated and regarded as scruffy, marginal, and indeed worthless by many non-desert dwellers; and, moreover, their inhabitants have been regarded in a similarly negative light. In an extraordinary combination of geography and social grade, historically the people living on or near them were frequently ascribed a low social status. Somehow the characteristics of the poor physical and economic environment were conferred upon their inhabitants by some, leading to the European deserts frequently being regarded as the poverty-stricken bad neighbourhoods of rural Europe. Many or most of these areas were long regarded as uncivilised and even lawless, and a refuge for the itinerant, indolent, social outcasts, outlaws and criminals. In the UK, they remain to this day destinations for social groups with nowhere else to go, including the homeless and travelling peoples of all kinds. In former times they were associated with the historic phenomenon of so-called squatter settlement too, when landless individuals and groups would seize plots of land to build houses and establish smallholdings, inevitably risking the opprobrium of existing settled resident populations. They were seen as not only worthless but ungodly too: for, historically, religious beliefs have held that such tracts of poor, bare soil and scrubby vegetation are 'against nature' – no part of God's bountiful gift of life on earth – perhaps making the people living on or near them inherently ungodly too. In some such places, non-conformist religionists often outnumbered adherents to the established church.

European deserts have historically been regarded as risky locations, places posing inherent risks to human health and wellbeing. And they are still considered risky by many in the twenty-first century, often with good reason. Historical attitudes towards these remote regions of Europe were very often negative. Before the nineteenth-century Romantics' glorification of the physical landscape and nature changed attitudes, the UK's 'untamed', 'wild' areas were feared by urban-dwellers terrified at the prospect of dangerous elemental forces, unpredictable and violent weather, dread diseases and threatening wildlife. Threats could come from their human populations too: in Spain, the parched backwater Bardenas Reales region in the central north-east of the country was long considered a dangerous haunt of bandits ready and willing to ambush unwitting travellers. While attitudes have shifted over time, some realities have not changed: the only UK species posing a genuine threat to the unwary is the venomous Adder, a native snake confined to the remaining desertic environments of the British countryside. Transmitted by the mosquito, malaria was endemic in many rural areas of Europe into the eighteenth and nineteenth centuries, most notably perhaps in the vast, remote Les Landes region of south-west France where surface geology and flat topography allowed surface water to persist year-round in some places, thereby maintaining the mosquito's breeding grounds and turning this vast former European desert into a disease-ridden hell for its unfortunate inhabitants. Even sand – a seemingly innocuous natural substance – has frequently posed an existential danger to human wellbeing and indeed survival, as strong winds mobilised dry sand, sweeping it across fields, roads, houses and whole villages. Farmland which had been carefully husbanded for decades might be swamped and sterilised on a large scale. A similar peril had been experienced at countless other European desertic locations for over millennia and, from the nineteenth century and earlier, a common solution to counter future such threats was applied in numerous locations across the subcontinent: tree planting on a massive scale. Intended to 'stabilise' precarious sandy soils and generally successful, the policy had, in addition, the cumulative effect that millions of hectares of formerly open European desert disappeared under imported conifer species. This invariably resulted in the near-total destruction of the former habitat and its associated biodiversity, with perhaps the only consolation being the introduction of what was often a new local economic activity – plantation forestry – to many rural backwaters.

Despite their acknowledged status as a distinct and unique semi-natural habitat type of considerable intrinsic merit, and their significant though reduced areal extent, the European deserts are neglected, routinely gaining little or no attention apart perhaps for occasional sensationalist media reports of fires breaking out in dry summers. These unglamorous environments attract little scientific attention, with several works by major ecology, landscape and countryside history authors omitting much or even any recognition of European deserts among chapters on woodland, wetland and other more conventional environments. They are absent from everyday media and discourse, even though some are located on urban fringes and many are popular with recreationalists. The volume of published material on them is scant and often difficult to locate, including the most basic data on location, extent, geology and geography. Their histories are not written, nor are the life and times of their people, and even in the second decade of the twenty-first century we do not fully know how these places were used in day-to-day terms. They are rarely celebrated in literature, perhaps because of their peripheral, scruffy and socially marginal status: Shakespeare set his ominously supernatural *Macbeth* Weird Sisters scene on a dangerous, uncontrolled 'blasted heath', but otherwise mentions are rare. Relatively few European deserts have engaged the attention of dedicated research communities, conservation societies, or charitable 'Friends' providing a watchful local eye and voluntary labour in the same way that more lauded habitats like mountains and woodlands do. Overall, little attention has been paid to them, and these unique areas have failed to attain any public profile even though their significance outstrips their collective areal extent.

The continued existence of many European deserts is threatened. Fragmented remnants are all that remain of many formerly extensive tracts of these rough lowland habitats. A grim line-up of external threats, historic and current, includes agricultural encroachment; poorly or unmanaged grazing; plantation forestry for timber production and carbon sequestration; sand and gravel extraction; annexation for military ranges and civil airports; all manner of domestic development including urbanisation, house building, new roads, outdoor sports and activities including off-road vehicle driving, straightforward trampling and general overuse; uncontrolled fire, invasive species, fly-tipping, air and water pollution and climate change. In practice, the degree of recognition afforded the European deserts varies markedly, and not all countries across the subcontinent allocate significant resources for their long term conservation, creating in some instances the perception of a neglected, undervalued and unimportant use of land. As the remaining tracts become degraded and further fragmented, the species which depend on these areas are threatened too.

MEANING AND SIGNIFICANCE

In aggregate the European deserts constitute a readily identifiable component of the environment of the subcontinent with a distinctive ecology. Manifesting also an extended history, they represent one of the cultural landscapes of Europe, and indeed offer a glimpse of an ancient European landscape. Resembling no other inland European environments, they hold a unique fascination, one transcending facts and objective accounts but also one which has been mostly uncelebrated by writers despite the manifest uniqueness of these places. They also constitute an element of the broader entity that is the global environment, and provide – for the imaginative – a tantalising glimpse of the seemingly endless plains, steppe and desert wildernesses of Africa, Araby, Asia, Australia and the Americas from the safe vantage point of the clement and green European continent.

Their continued existence is a testament to centuries-old traditional land management techniques – which in turn were adapted over time to suit changed circumstances and a fluctuating climate – ingenuity, energy, doggedness and grit, community enterprise and in many cases benevolent legal structures enabling poor village-dwellers to win a livelihood in the face of large landowners. Their stories will not always be happy ones – thousands will have struggled to survive on poor soils and amidst carefully crafted legal restrictions; and many will have lost the battle against rural starvation over the centuries as former commons were fragmented piecemeal or extinguished wholesale through enclosure by powerful interests, with former ancient rights to access, usage and materials there terminated.

The ethos and content of this book are based primarily upon scientific, rational and objective ideas. But the European deserts have meaning and significance over and above these dimensions. In seeking to understand and appreciate these and similar places, our cumulative experience and learning derive partially from a synthesis of the cognitive and the affective. We experience and learn about the world, and develop attitudes towards it, primarily in an objective and cognitive-based manner, involving facts and knowledge. But we also employ affective mental processes influenced by emotions, feelings and beliefs to simultaneously experience the world through a subjective lens. The affective brings a subjectively different dimension to understanding and appreciation, and brings the power of the imagination to be engaged in the production of perceptions. The Romantic artistic movement of late eighteenth-century Europe challenged the untrammelled elevation of objective, scientific and rational ideas belonging to the Age of Enlightenment with the power of individual imagination and subjective experience. Nature was a central theme and driver of the Romantics, with poets and painters demonstrating their appreciation of landscapes, and especially wild or sublime scenery, counterposed with a rejection of industrialisation and the then-new, crude, human-made vistas of the Industrial Revolution. A central concept was that humanity's true self may be found in the wilderness rather than in the city. These places are almost as significant for what they represent as what they are – and enjoying fresh air and light breezes over an open tract of European wilderness under clear blue skies offers a glimpse of the planet in the raw, amidst a serenity of peace and quiet.

EUROPEAN DESERTS: TOWARDS A WORKING DESCRIPTION

To commence this exploration of some of Europe's most interesting dusty backroad wildernesses, a working description of the continent's deserts is necessary. However, no suitable and comprehensive working description currently exists, a fact unsurprising because these places have never been identified or examined in such a collective fashion. The term desert is here employed in a manner that supports the intentions and spirit of this work, emphasising their broad environmental and habitat characteristics. Any working description needs to be couched in broad terms, because while exhibiting sufficient similarities, the deserts of Europe are nonetheless diverse in terms of their origins, landforms, soils, fauna, flora, water balances and human histories. Because of this diversity, a wholly objective definition of these environments is not essayed here, but instead a broad and necessarily subjective framework for identification has been derived. A set of fourteen indicative criteria will in combination assist with the identification of instances of this distinctive environment type across the length and breadth of Europe. Those criteria are as follows:

1. Heathland as European desert – the habitat of the tracts styled European deserts for the purposes of this book is conventionally referred to in broad terms as lowland dry European heath. Relatively uniform from a physiognomic point of view at the landscape level, in common usage heathland means any kind of barren waste ground, especially where trees are sparse or absent and the prevailing landscape is largely bleak, bare and open. Strictly, heathland is vegetation dominated by dwarf shrubs, but the term is conventionally applied to a range of habitats with common characteristics, transcending several habitat classifications which are closely ecologically related and exhibit intergradation. These other communities include humid and wet heath, dunes, scrub, acidic and other grassland and bare and unvegetated ground.

2. Location and geography – the places in focus here are:
    - located on the European subcontinent
    - occur in most European countries, with a concentration across the western and north-west margins of the subcontinent – and in particular:
        - the majority of the deserts are located throughout an arc sweeping from Atlantic Europe to central Europe, from Portugal and Spain in the south-west through France, the British Isles, and across northern Europe including Scandinavia to the Germany–Poland border
        - within this broad picture, there is a dense concentration in Belgium, the Netherlands and Germany
        - there are fewer instances in central and eastern Europe, including countries from Poland in the north to Bulgaria in the south
        - an outlier instance, Iceland has extensive lowland and alpine boreal heaths set among the large, bare areas of open wilderness that characterise the island
    - located in lowland regions – at moderate altitudes up to 250–300 m, clearly distinguishing them from the upland moorland dominant over some of the subcontinent's hill and mountain areas (though several instances are set at higher altitudes, including in Spain where the European dry heath habitat is recorded at up to 1,900 m)
    - located inland – mostly distinct from coastal sites, and often in relatively remote regions of Europe; and
    - markedly varied in size: most are small, and exist as mere fragments of their former extent.

3. Climate – the deserts are subject to a predominantly temperate maritime annual climatic regime, one which is warm and wet in general terms and varying, to the east and south respectively, with continental and Mediterranean influences.

4. Relief and geomorphology – their appearance is variable, but the landforms of the European deserts are generally moderate in appearance, varying from lowland flats and plains to plateaus and low hills. Ridges, notably in the Netherlands, are often glacially derived. Some of the Spanish deserts exhibit flat-topped mesas more usually associated with the arid zones of the USA.

5. Geology – distributed over a range of geological substrates, though perhaps best termed mineral soils in general, including sand, gravel and glacial outwash through coversand to the alluvial and marine-derived rock geological substrates of the Spanish Tabernas semi-desert; elsewhere, sometimes with peat.

6. Hydrology – these are free-draining and relatively dry environments, though local impermeable subsurface geological conditions often result in small-scale aqueous habitats including ponds, marsh and bog and/or springs and streams which may be ephemeral. In many instances, the water table lies far below the ground surface.

7. Soils – typically sited on impoverished, nutrient-poor and frequently acidic mineral substrates, many soils are fully weathered, highly permeable through which precipitation readily sinks and of relatively low nutrient status. Occasionally with areas of bare soil, sand, or rock.

8. Ecology – European deserts host many threatened species on the sparsely vegetated habitats that have developed over nutrient-poor soils. They typically support a limited range of dwarf-shrub vegetation dominated by heather species – which are more usually associated with upland locations – and gorse, with relatively few higher plants. In a few locations they are barely or un-vegetated, which is exceptional in verdant Europe. The faunal profile range is relatively limited, too, though with characteristic birds, reptiles, and a diverse suite of invertebrates, in Britain providing home for up to 40% of the country's spider species. Most of these habitats are secondary in character, dependent on human intervention, particularly grazing and fire.

9. Landscape – this is a key characteristic distinguishing the European deserts, their terrain presenting a striking visual contrast to that of their frequently verdant and civilised surroundings. Typically flat or lightly sloping, many are open in aspect, with few or no trees making for extensive and unimpeded though mostly unspectacular views. The occasional stream or small river valley may cut through to form sometimes unexpectedly steep sand or rock inland cliffs. Dry, dusty, sparsely vegetated and relatively scruffy or untidy to the eye, many European deserts share the visual characteristics of wilderness, offering a prospect of abandonment, dereliction, desolation and often otherworldliness. Some areas founded on drift sands develop large wind-blown deflation (wind-eroded) ground features, extraordinary areas of bare and actively moving sand set between often substantial remnant plateau dunes (called 'forts' in the Netherlands). Many

European deserts are predominantly unfenced and unenclosed, too, offering uninterrupted landscape vistas and lacking the highly visible fences, hedges and other field boundaries enclosing the surrounding farmed countryside.

10. Economic – while often surrounded by arable fields, these are usually sites of poor-quality land in agricultural terms, whose severe limitations make them unsuited for growing crops. Some are used for low-intensity grazing, some are planted with trees in industrial plantation forestry, and some provide environments for public recreation, typically serving nearby population centres. But while most of these places have little or no significant economic role, in many instances their twenty-first century economic redundancy contrasts sharply with their historical role: in some places, until about 100 years ago they were the foundation of sophisticated and intensive regional agricultural farming regimes.

11. Formal terminology and land classification – some are legally designated as nature reserves, nature parks, national parks and similar. Many are actively managed for conservation objectives, though many are not – designation alone does not necessarily imply active habitat management. The European Red List of Habitats identifies 'Heathland and scrub' as a specific terrestrial habitat distributed across the subcontinent, comprising 40 habitat types dominated by assemblages of woody shrubs often in combination with herbs, and sometimes with mosses, liverworts and lichens. In the UK, lowland dry acid grassland and lowland heathland are specified as national Priority Habitats for conservation. Inland sand dune habitat is included, though it is scarce, being found only in the Breckland area of East Anglia (Chapter Seven).

12. Naming – these types of location are referred to with a range of terms across Europe. A brief review of those relevant includes ager, badland, bos, common, *desierto*, forest, heath, heide, *landes*, mark, moor, muir, *pustynia*, *saltus* (Latin for wild, uncultivated land), steppe, weald, wald, walk, warren, waste and wold among others.

13. Vulnerable – most or all were much more extensive in former times, and despite their distinctive ecological values the continued existence of many European deserts remains under threat. Presently covering in aggregate a small proportion of the terrestrial face of Europe, many of these areas are declining in both areal extent and ecological value. The European Red List of Habitats recognises dry heath to be a vulnerable habitat, with identified threats including agricultural expansion and intensification, afforestation, infrastructure and housing development, wildfires, rural abandonment and the cessation of traditional land uses and environmental management in broader terms.

14. Sociocultural – the deserts of Europe typically lack any significant settlement, houses, or even highways, and are usually unpopulated. Yet while many or most existed for millennia before humanity's influence became significant, they may be referred to as cultural landscapes because humans frequently played a decisive role in their creation and have subsequently maintained many since the last Ice Age through purposeful management. Without human intervention over millennia,

most would have succeeded to woodland. Cultural prejudices have led to these locations regularly being regarded as low-status places with historic associations to poverty, social marginalisation, lawlessness and risk. They are largely unknown too: the deserts of Europe are obscure, uncelebrated, disdained and only infrequently do they figure in the narratives of national histories.

This book will identify, appraise and celebrate a selection of some of the most interesting and distinctive desert-like environments on the European subcontinent. It does not venture to undertake a comprehensive survey of the European deserts.

The selection of individual instances for each of the country chapters has been undertaken on a pragmatic basis, driven primarily by the availability of scientific and other literature on these mostly little-known places, and also the author's personal knowledge. Though largely unsung and generally unspectacular, they are distinctive and fascinating environments, offering a rare glimpse of the planet in the raw. While as worthy in their way of stewardship as any of Europe's grand palaces and treasured landscapes, they have little or no public profile and have frequently been neglected over past decades, both individually and in aggregate. Just as drylands exhibit great diversity at the global scale, the deserts of Europe vary significantly across the subcontinent. Nonetheless, they are as a group distinct from other types of semi-natural environments, and exhibit sufficient shared characteristics to constitute a distinct and well-defined study population. That they are comparable is reinforced by the fact that human populations living on or near these unpropitious lowland wildernesses with relatively few conventional resources spread over the four corners of Europe have faced similar existential challenges which they have addressed in similar ways. People in these places have frequently developed similar approaches and adaptations to farming and environmental management to survive and thrive that are distinct from those used in more favourable farming environments. Some, including large mammal herbivory – extensive livestock grazing – may be recognised as similar to those employed in arid lands across the planet. Europe's deserts evince many parallels with those dominating more than 40% of the world's land surface, and the present work represents perhaps the first recognition and exposition of that fact.

The European deserts play a crucial role in the traditional landscape of lowland Europe, and are recognised for the value of their semi-natural environments and also their cultural heritage, among other attributes. However, precise data on the number and extent of these tracts are lacking. The number of individual sites scattered across the subcontinent is estimated to run into the tens of thousands, though an exact figure is unavailable. Reliable data on the total extent of the deserts of lowland Europe is similarly lacking, though it is estimated that they account for less than 1% of the total terrestrial area of the European subcontinent.

COUNTRY CHAPTERS

The country chapters commence at Chapter Two with Spain, home to two of the most extraordinary examples of European deserts. Chapter Three focuses upon the Netherlands, a small and intensively developed and highly managed country with relatively large areas

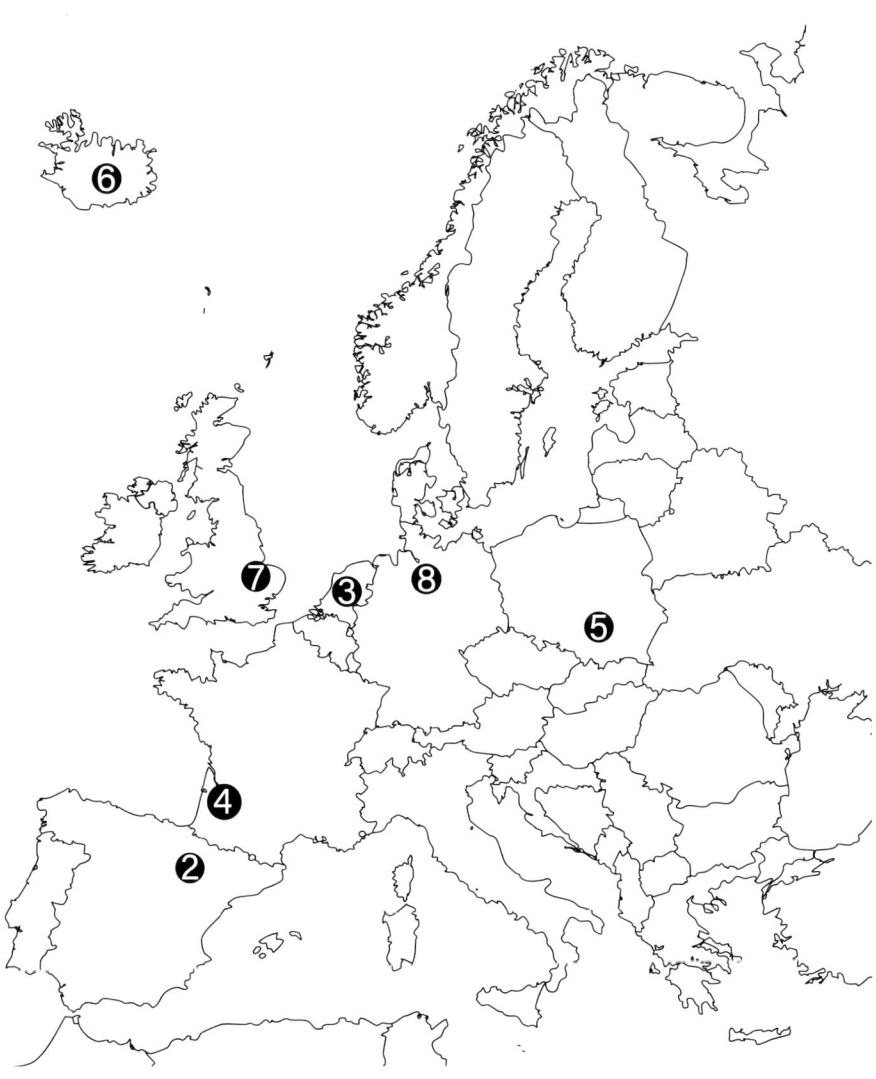

European deserts: sandy wildernesses. (Montserrat Morillas)

Chapter 2 Bardenas Reales, Spain
Chapter 3 Kootwijkerzand, the Netherlands
Chapter 4 Les Landes, France
Chapter 5 Pustynia Błędowska, Poland
Chapter 6 Ódáðahraun, Iceland
Chapter 7 Breckland, United Kingdom
Chapter 8 Lüneburg Heide, Germany

of dry sand wildernesses and one of the few in Europe to expend significant resources upon their conservation. Chapter Four is set in France, examining what was formerly perhaps one of the largest in Europe, the Les Landes region of Aquitaine. While this south-western France desert region has been almost completely stabilised – planted with trees – the desert is still there if you look for it, and has an extraordinary history to tell, one illustrating the poverty yet also the majesty of these environments. Chapter Five looks at a remarkable instance in Poland, where miners created a desert and where mirages more commonly associated with the Sahara have been reported! Iceland is the focus of Chapter Six, a 'desert island' in the North Atlantic and the largest and probably the coldest desert in Europe. Chapter Seven presents an investigation of an English European desert, the little-known Breckland district of eastern England. Chapter Eight focuses on a German example, an exceptional instance in a country that once had perhaps the greatest expanse of this habitat in Europe. The final chapter concludes the book, by:

- offering summative analyses and insights into the phenomenon of the European deserts
- identifying extraordinary desertification processes presently in progress in south-eastern Europe which offer insights into the historical formation of the subcontinent's deserts
- considering Europe's cold deserts in the context of the contested Anthropocene thesis and global processes of climate change
- adopting a novel approach, presenting an outline of the author's desired vision for the European deserts in the year 2050.

Let us look at several notable examples of the dwindling cold deserts of Europe, these mysterious, enigmatic, yet fascinating, romantic and priceless survivals whose continued existence is significant at the local, subcontinental and indeed global scales for their distinctive environments, their biodiversity – and their place in the history of humankind.

Chapter Two

SPAIN: BARDENAS REALES – IBERIAN BADLANDS

A big village with a population of 2,300 residents, Arguedas lies sweltering in northern Spain not too far south of the Pyrenees in the centre of the Ebro basin, astride the road to nowhere in particular. Streets here are narrow, and little traffic disturbs the cracked tarmac or the *señores* nursing coffees while sheltering from the intense sun at the Bar Moderno. The dominant colours in built-up Arguedas are brown, beige, buff and… dust, with plenty of whitewash too in that southern European way, the intention being to reflect as much heat as possible from *las residencias*. Here, we are a long way from the sophisticated fleshpots of Spain's northern coast, with the upmarket beaches, international art galleries, and museums of San Sebastian and Bilbao. Like so many other small towns here in the Navarre region, an awful lot of houses and several shops along the main thoroughfare have clearly not seen much traffic in recent years either, being boarded up. Gaps in the narrow street's frontage mutely signal where others were demolished, and the handsome young pavement trees gamely planted along the central Calle Real fail to achieve much of their intended impact of softening up the bare streetscape. Determining to stay in flaming Spain at the height of the torrid Iberian summer for research purposes was adventurous – or perhaps simply foolhardy – and the tatty bedroom air-conditioning unit proved essential in the local hostelry beside the dusty road out of town where watching bullfights on the barroom TV was the main recreation for patrons.

Arguedas lies squeezed between a dramatic line of inland cliffs and the flat, wide and fertile valley of the river Ebro, which is known – somewhat improbably – for the production of a variety of rice called lido, which must inevitably require an awful lot of water to grow in what is a dry region. The parched yellowy-white cliffs rearing up immediately behind the village formerly contributed to the local stock of residential accommodation, an equally unlikely fact given away by a spooky series of domestic door-sized rectangular holes located along one specific horizontal cliff stratum. The caves of Arguedas were dug towards the end of the nineteenth century as homes for local people unable to afford more conventional dwellings, or – so the story goes – those seeking to avoid having to live at the whim of the few local landlords who controlled most of the town's houses. The relatively soft geology here meant that a simple pick and shovel coupled with a lot of effort could produce a family home, one which could easily be expanded with more rooms as the family expanded, and within which the animals were often kept too. Another

Former cave houses in Arguedas.

major advantage of these bioclimatic shelters is that they remain at a steady temperature of 18–22 °C year-round, enabling their inhabitants to withstand the dramatic annual temperature range outdoors without any need for electromechanical cooling devices, and bringing to mind the cave dwellings of Petra in the roasting Jordanian desert. By 1940, over 50 cave homes had been cut into the bluff inland cliffs here and, doubtless, generations of Spanish and Basque families grew up perfectly happily there overlooking the settlement below, however low-status their rock houses – *cuevas*, or, literally, caves – might have been perceived. The *cuevas* were gradually abandoned in the 1960s with the construction of social housing – described as more dignified homes – and while most are now blocked off, a few have been repurposed as tourist accommodation. However, the cultural practice of cave-dwelling is not at all unusual even in twenty-first-century Spain: the town of Guadix in Andalucía has roughly 2,000 underground houses in occupation, the cave district being signposted from the main street as the Barrio Troglodyte.

But nothing in this sweet though parched village prepares the casual adventurer for the incomparable experience waiting quietly, unassumingly, just a few kilometres to the east. Arguedas is the gateway to the most extraordinary natural panoramas in Europe: Bardenas Reales, a European desert if ever there was one. Take the lonely road rising east out of town, marvelling at how quickly views of the generally green and fertile river valley so rapidly change to those of a lightly farmed, treeless and open landscape. Here you will head into a desertic steppe environment; it's like going directly from one climatic zone type into a completely different one, yet over a distance of a mere ten kilometres. With the surroundings growing drier and drier, browner and browner, the adventurer is slowly but finally greeted by a sweeping and frankly staggering semi-desert landscape like nowhere else on the subcontinent. In sharp contrast to the otherwise generally lush, green landscapes of northern Navarre surrounding it, one immediately senses the dominance of the physical environment here. The geology is derived from a former palustrine basin – brackish water – situated here 20 million to 10 million years ago, when a horizontal series of alternating hard and soft rocks were laid down in a classic lake depositional sequence. These silts, chalk, gypsum, sandstone and limestone were subsequently sculpted by erosion into today's spectacular landscape of bone-dry gullies, canyons, flat-topped plateaus, solitary hills with precipitous vertical flanks, badlands, salt deposits and sunburned cracked clay, a genuine semi-desert writ on what is a huge scale for dinky old Europe. It is also a biodiversity location of international significance, hugely important for conservation and heavy with legal designations including as a UNESCO Biosphere Reserve, a Natural Park, a Special Protection Area for its bird interest, and a Special Area of Conservation for the habitats and species there, including insect-eating shrews (upon which the Ladder Snake preys) and twelve species of snail, which form an important source of protein in the food chain here.

BARDENAS REALES ENVIRONMENT

Bardenas Reales is a large erosional depression framed by heavily dissected plateaus to the north and south and subject to a climate of extremes. The rock-desert landscape featuring extensive badlands is conventionally divided into three geological geomorphological zones: to the north and west lies a great plateau known as El Plano, almost completely flat and with large areas of arable fields dispersed among tracts of dry, rocky land deemed too barren to attempt cultivation. To the south is La Negra, a series of plains formed of limestone and sandstone plateaus with dramatic rock features intruding into the landscape here and there. But the central area, La Blanca or Bardena Blanca, offers the greatest desertic spectacle, at 25,000 ha a vast and mostly unvegetated area of naked sedimentary rock, appearing dried-out, liberally dusted with salt, and dazzling in the sun. Here, the scarcity of more resistant rocks capable of protecting the clays and silts has resulted in La Blanca being the most eroded section, appearing instantly recognisable as a desert in Europe.

The horizon hoves into view in a raw-earth spectacle shimmering gently in the haze. Here, the heat hangs heavy at midsummer, amplifying the perception of a quietly extreme environment. Unimpeded by trees or buildings, vistas here are epic, fading into a hazy distance of blurred brown, buff, white and grey under huge, bolt-blue skies. The calm is gently pervasive and profound, the silence remarkable and also somehow relentless. At midday, absolutely nothing moves. The air seems thin. The light drenches the landscape today just as the infamous fifteenth-century Bardenas bandit Sanchicorrota would have once witnessed it, amplified a hundred-fold by the harshly reflective light-hued geology here. Together, these physical phenomena transfix the puny human, marvelling at a scene which seems like some sort of suspended animation – located in an oven. It is not hard to conclude that humanity has no place here. Rarely does one get so super-close to the naked, raw earth in such an extreme setting: live, living geology and geomorphology, and their associated biological consequences. Most of Europe isn't like this... but, clearly, it can be.

The symbol of this remote and little-known Spanish natural park is Castildetierra, an isolated pinnacle landform rearing up from the Blanca plain like some sort of accusatory finger pointing at the dry, dry, endlessly blue sky. Meaning 'Castle made of earth', Castildetierra is an earth pillar, a narrow column of honey-coloured unconsolidated material left over from thousands of years of erosion by wind and water and rising with a haughty swagger atop a small island-like pyramidal landform, tapering upwards and capped by a small layer of resistant rock sheltering the friable column of clay and earth layers stacked uneasily below. In its way, Castildetierra embodies many of the best features of Bardenas Reales in a bite-size, quick-visit-friendly format. In France, these concretion pedestal formations are termed *demoiselles coiffées* – damsels with hairdos – and in the USA they are hoodoos, though I think I prefer the French. As time progresses, however, continuing erosion of the soft lower strata is gradually undermining the cap, which will eventually tumble leaving the remnants of the pillar to be quickly eroded. In a very visible microcosm of the processes happening around it, poor old Castildetierra is even now gently declining in height and will eventually be no more. Parts of the landscape

Semi-desert landscape of Bardenas Reales Biosphere Reserve, Navarra.

An isolated earth pillar, Castildetierra is the most iconic image of the Bardenas Reales Parque Natural y Reserve de la Biosfera.

around this iconic feature feel as desertic as anything in Africa or Asia, and some are visually comparable to the stark, heavily dissected tablelands of Monument Valley on the Colorado Plateau, whose 1,300 ha have defined the image of the American West cherished by decades of moviegoers. Bardenas Reales attracts makers of TV commercials and films seeking otherworldly backdrops: one of the James Bond spy-thriller films *The World Is Not Enough* was shot here, as was season six of the drama *Game of Thrones*. But more usually the place is quiet: very, very quiet...

Recognised as a unique wilderness for centuries, Bardenas Reales extends over an elongated 45 × 24 km tract of empty land. A few small roads cross the area, but some terminate unexpectedly to become bumpy, white, dusty tracks strewn with rocks, along which the casual visitor is not recommended to drive after heavy rain. But, really, it's best to abandon your vehicle and walk (or bike) through this extraordinary place to experience the parched Bardenas full-on. A consequence of geology and climate, the area is subject to an intensely erosive climatic regime, with substantial surface and subsurface water erosion. Unusually for Europe, wind erosion also plays a major part in sculpting the unique, abrupt landscapes, canyons, plateaus and precipitous edges there. Like the *mistral* of the Rhone Valley, the *cierzo* blows over central northern Spain from the north and north-west, a powerful cold and dry wind with substantial erosive and desiccating power which is partially responsible for the dramatic landscapes. The *cierzo* attains speeds of 100 km/h or more for over one-third of the year through autumn and winter, lofting any loose soil, and in July 1956 a barely credible maximum velocity of 160 km/h was recorded. This wind is known from ancient times, when Roman senator and historian Cato the Elder in the second century BCE described it as one that fills the mouth and tumbles waggons and armed men; in the twenty-first century light trucks traversing the exposed regional roads get knocked about when the wind is in full howl. Its drying qualities enhance the aridity of the Bardenas in addition to animating the hundreds of wind turbines located around the reserve, and the many hectares of solar panels in this relentlessly sunny region also contribute to Navarre's electricity supply, two-thirds of which is from renewables. Here, the semi-desert weather climate is an economic asset, in the forms of wind and solar energy. Sparse vegetation contributes further to shaping the terrain, as what precious soil exists there is exposed to the elements.

Set at altitudes of 280–660 m, Bardenas Reales is located in central northern Spain, a country of dramatic landscapes and natural features. The desert's altitude exceeds the lowland altitudinal maxima of 250–300 m set out in the guideline descriptive criteria for European deserts in Chapter One, but the terms of reference are different in Spain, a country dominated by mountain chains and plateaus and with the distinction of being set at an average elevation of 660 m, the highest of any western European country. And no appraisal of Europe's most desertic tracts would sensibly exclude the Bardenas, a unique place whose scale and climatic and landscape drama is unrivalled. Here are dramatic physical features including mesas and buttes, more usually associated with desert regions on faraway continents. The word *mesa* means table in Spanish, and well describes these broad monolithic features standing proud above the Bardenas countryside as a table stands above the kitchen floor. Vistas of mesa landscapes are dominated by the horizontal. These wide, flat-topped, and frequently isolated hills are formed of horizontally bedded sedimentary rocks. They are capped with a thick slab of resistant rock over relatively soft strata, with their geology frequently very visible on their steep and largely unvegetated flanks. A butte is a tall, steep-sided and narrow rock tower, typically isolated and sticking dramatically out of plain landscapes. A geomorphologist's rule of thumb holds that while the butte has a top narrower than its height, the width of the top of the mesa exceeds its height. Buttes were created through processes of erosion and weathering, the gradual wearing away of the earth by wind, water and ice to leave perpendicular sides. Often a resistant cap rock with a flat top overlies layers of less resistant materials. While buttes are distributed globally, with those in the USA perhaps the best known, they are exceedingly unusual in tame Europe; nonetheless, they are features of the Bardenas, as are mesas.

Bardenas Reales is located in one of the world's arid zones, dominated by a true desert climate characterised by extremes. While classified under the Köppen climate system as BWk – a cold arid climate – everything is relative: summers here are unfailingly hot and dry and rainfall is scant, irregular, and yet frequently torrential. With maxima in spring and autumn, the rainfall is equinoctial in nature, making summers seem yet more harsh in their dryness. In common with other world arid zones, the winter season is dry, in contrast to the wet winter more characteristic of the Mediterranean climatic regime

Bardenas Reales mesas.

prevailing elsewhere in Spain. Lows of 5 °C are typical, and while snow is rare it is not unknown in this region of extremes. La Blanca is the driest zone, where an annual average of 300 mm of precipitation is dwarfed by potential evapotranspiration of over 1,100 mm. Rainfall in the wider Bardenas territory is highly variable between years: some years are wetter, some are drier, with a large interannual variation from as little as 190 mm up to 850 mm. Much of the erosion here is triggered by slope wash during torrential rain events, whose substantial erosive power produces rapid surface runoff. Over areas where soil and rock materials are easily weathered and vegetation is sparse or absent, loose surfaces readily slump and slide downslope. Little water infiltrates into the soil at such times, meaning that most of the precious liquid is unavailable to plants. Similar erosion may occur when winter snow cover melts, too. No rivers cross this harsh desert, and the few watercourses are seasonal, remaining dry for most of the year, and their runoff waters tend to be of low quality due to the high salt content. Surface water is almost absent apart from a few springs to the north and a couple of human-made reservoirs, but there is just enough to sustain populations of the European Pond Terrapin, a near-threatened species once common in northern Europe but now rare in most countries and largely confined to the south of the subcontinent.

The most dramatic climatic variable here in northern Spain is temperature. July is the hottest month, when temperatures frequently attain 35 °C for several days at a stretch. But the all-time record high in Bardenas Reales was 44.5 °C on 7 July 2015. At such extreme temperatures plants experience severe physiological problems – and humans extreme discomfort. On that day, with the weather coming in from super-hot, hyper-arid north Africa, temperatures in northern Spain approached the highest ever recorded in the Sahara, when in July 2018 a reading of 51.3 °C was registered in the small Algerian town of Ouargla. An important insight to draw from this is that the climatic zones so precisely delineated on climatologists' maps are only ever approximations of messy reality: desert climate conditions may and do prevail in tame, temperate Europe, however infrequently.

BADLANDS

Badlands are naturally occurring morphological features characterised by a dominance of erosion processes and steep topographic gradients. These surreal, fantastical terrains fascinate geologists, geomorphologists and the casual visitor alike. While very uncommon in Europe, they are nonetheless distributed around the planet, largely but not exclusively located in semi-arid regions, and are frequently desert-like in appearance. Badlands present bold vistas way beyond the everyday European landscape experience, and while often described as otherworldly – looking like another planet – such interstellar comparisons are unnecessary: here, you are gazing upon processes and landforms more characteristic of the arid desert regions of North Africa. The word barren really means something in this place of extremes. Vegetation is sparse or absent. Stark-naked clayey slopes are steeply inclined, their geology plainly visible in the absence of soil. Knife-edge ridges abound in an intensely dissected erosional landscape. A dense, hierarchical drainage network of deeply incised hillslope gullies with V-shaped profiles produced by the erosive action of flowing water all head giddily downwards. Rapidly evolving, these precipitous drainage systems carved into such steep slopes make for a dramatic landscape, and the characteristic recursive pyramidal slumping slope edge forms look like some sort of statistical algorithmic bifurcation diagram! Washes dominate on lower slopes; on the plains below, some areas are strewn with rock debris while others look like melted cheese.

Over most of the planet, the natural erosion processes which form today's familiar, mature and fertile landscapes are normally imperceptible day to day. In Europe, most of the land surface ceased eroding appreciably a long time ago, and the 'good-lands' we recognise in regular rural areas are mostly set in fully weathered landscapes, with long-established soils, relatively sparse drainage systems, and their physical features mostly cloaked in agricultural crops, woodlands, or water features. But much of the Bardenas badlands are youthful in geomorphological terms, and in comparison to the 'good-lands' elsewhere appear dramatically different – naked exposures of the planet in the raw, visibly chaotic and very obviously an active geomorphological work-in-progress rather than in any sense a settled, mature landscape. Continually raging weathering processes dominated by torrential overland flow often produce rapid erosion rates, and offer few or no prospects for successful plant establishment. Several erosion processes occur concurrently in the Bardenas badlands, including piping, where clays and other sediments contract at the surface facilitating water infiltration, and the subsequent development of shallow underground pipes leads to subsurface tunnel erosion. This then may evolve into chains of sinkholes at the surface through roof collapse, resulting in new gullies producing

SPAIN: BARDENAS REALES – IBERIAN BADLANDS

Heavily eroded Bardenas Reales hillside.

Bardenas Reales unstable hillside mass wasting.

Dramatically eroded Bardenas Reales landscape. (Mikipons, CC BY-SA 3.0 ES)

Bardenas Reales badlands panorama.

treacherous landscapes riddled with holes and pinnacles and blighted by landslides. The rapid geomorphological evolution proceeding in the Bardenas means that many of the processes and landforms normally occurring at a snail's pace in most landscapes may be observed here, speeded up and on a miniature scale.

Badland landscapes are perhaps best known in the USA. The badlands of the Cheyenne and White Rivers region of South Dakota were avoided by early settlers heading west, with eighteenth-century French-Canadian trappers damning the region as '*Le mauvaises terres a traverser*',[1] meaning 'Bad lands to go through'. In these badlands, erosion proceeds at an average rate of 25 mm annually – a positively turbo-charged rate in the sedate world of geology, and especially when compared to the granite of the nearby Black Hills, where erosive rates of 25 mm *per 10,000 years* are more the norm. But what is so bad about these lands? Frequently described – and employing a clear value judgement – as severely degraded, badlands everywhere are disdained as being outside the realm of conventional productivist conceptions of land use, and the term goes hand in hand with images of poverty. In Europe, where high agricultural output has long been a goal of individuals and governments alike, badlands are a godless phenomenon: no part of God's bountiful gift of life on earth, they are against nature and against economics. Despite centuries of advances in agricultural techniques, significant agricultural use is usually impossible on these dynamic areas of bare soft rock terrain which offer plants no permanent foothold. Unlike other, less extreme European locations where remaining fragments of undeveloped land may be brought into agricultural production, few measures are available to bring badlands into the twenty-first-century productivist fold. Attempts at settlement are usually equally futile, and in evincing minimal or no observable human activity some sections of the Bardenas seem cut off from the everyday world. *Good.* These thrilling and impossibly dramatic landscapes are part of the fascination of Bardenas, offering a near-unique spectacle rivalling that of Death Valley in its breathtaking but also somehow sinister nature – the casual visitor here readily perceives a sense of personal risk. The badlands of Bardenas Reales are an authentic, gritty, grubby, stark, parched desert landscape located on the green subcontinent, and there is little else like them.

Bardenas Reales is at least partially a cultural landscape, the result of the centuries-old practice of itinerant sheep grazing as well as the specific climatic and geological conditions, and all three factors are reflected in the biodiversity. The ecosystem here is a fine example of the Iberian steppe, and comparable to those of the semi-arid steppe regions of North Africa or Central Asia: it is similarly treeless but with a yet more arid climate, and species of halophile (salt-loving) flora. More precisely termed pseudo-steppe, it is a pioneer or semi-pioneer community type with a Mediterranean distribution that resembles steppe but is in fact desertified. Scarce in Europe, pseudo-steppe is classified as a priority habitat under the European Commission Habitats Directive on the conservation of natural habitats and fauna and flora. An open landscape dominated by a

---

1   Froiland, S.G. (1990) *Natural History of the Black Hills and Badlands.* Sioux Falls: The Centre for Western Studies, p. 177.

Bardenas Reales: heavily eroded triangular peak.

herbaceous layer, pseudo-steppe is a semi-natural agrarian system with a high ecological value dependent upon sheep grazing for its long-term maintenance. Other habitats here include heath, sclerophyllous scrub – flora adapted to the long periods of dryness and heat of the Mediterranean summer – and thinly vegetated grassland. Though the available herbage is limited and scattered, sheep grazing is essential: too low an intensity results in scrub encroachment, biodiversity decline and an increase in the risk of fire, while overgrazing may accelerate desertification.

What about wildlife in this arid wilderness? Numerous protected animal species depend, to a greater or lesser degree, on this habitat type, and with the ability to withstand the extreme conditions, invertebrates are an important basis of the food chains in the steppic areas here – dragonflies, mantises, grasshoppers, beetles, butterflies and spiders. Larger mammals include Wild Boar, Roe Deer and Rabbit – in fact, the Rabbit is considered a fundamental part of the Bardenas ecosystem, and is the traditional hunting species here alongside hare. Myxomatosis hit in the 1950s, and another virus causing haemorrhagic disease in 1989 contributed to another major die-off of wild rabbits, in turn causing a decline in the populations of species that would normally predate them. In La Negra, dominated by Mediterranean scrub, there are Red Fox, Wildcat, Genet, amphibians and reptiles. Conservation of pseudo-steppe is very important for several steppic birds, some of them globally endangered. The avifauna community comprises over 100 species, with two significant groups being birds of prey and birds of the steppe. Twenty-four species of raptor include the globally threatened Egyptian Vulture, Griffon Vulture, Eagle Owl

Black-bellied Sandgrouse inhabit semi-arid plains with steppe to semi-desert vegetation, often feeding on small seeds, grain and cultivated legumes. (ramidos, CC BY 4.0)

Dupont's Lark – a bird of open sandy semi-desert and steppe areas with some grass, feeding upon seeds and insects. (Wim de Groot)

EUROPEAN DESERTS

Endemic to south-western Europe, the Ladder Snake preys upon small mammals, spiders, insects and birds, preferring habitats of stones, boulders and low shade. (FOTO-ARDEIDAS, CC BY-SA 3.0)

Egyptian Vulture in flight. (PJeganathan, CC BY-SA 4.0)

Wild Boar are omnivores able to adapt to several habitat types with shrub cover – including savanna – and a suitable food supply. (FOTO-ARDEIDAS, CC BY-SA 3.0)

Esparto Grass, a perennial of NW Africa and the Iberian Peninsula, has been managed by humans for centuries for its fibre which has great strength and flexibility. (David Elliott, CC BY 2.0)

and Golden Eagle rock birds. The Iberian Peninsula is the most important region for steppe birds within the EU, and many steppe-land bird species are associated with traditional agricultural practices. Steppe birds here include the resident Great Bustard; with adult males weighing in at up to 19 kg, this ground-nesting bird is among the heaviest living flying animals. The International Union for Conservation of Nature's Red List of Threatened Species classifies the species as vulnerable to habitat loss, making a historic description of the cooked flesh of this stately game bird as 'delicious' rather unfortunate. Here the Great Bustard is at the western edge of its range, which extends across Central Asia. Notable other birds include Little Bustard, Stone-curlew, Black-bellied Sandgrouse and the shy and scarce Dupont's Lark, whose European range is confined to Spain. Large areas of the territory are off-limits to human visitors from March to September to minimise disturbance of ground-nesting birds. With intense desiccation in summer and winter, drought-resistant species are favoured, and steppe vegetation includes Esparto grass, a perennial grey-green needlegrass growing over the Iberian Peninsula and North Africa which produces a fibre of such strength and flexibility that it was for long used in the soles of espadrille shoes – once the footwear of peasants – and in ropes, baskets, mats and paper.

AGRICULTURE

Despite some extremely challenging soil and climatic conditions, agriculture has long been practised in this European desert. One translation of the word Bardenas is 'Low lands where the sheep graze' (other translations are available), and extensive grazing dominated over Bardenas Reales for many centuries. Today perhaps 90,000 to 120,000 sheep are moved into the area each year after a summer in the Pyrenees to take advantage of more clement pastures for the winter, following the age-old tradition termed transhumance. Bardenas Reales lies on a traditional transhumance route, one of a network of drovers' tracks called *cañadas reales* – royal livestock trails – along which livestock are legally permitted to transit and to graze. Mostly oriented north–south and most winding tortuously through empty central Spain, where the annual weather regime is regularly described as nine months of winter and three months of hell, the longest is 800 km in length, which must have been an epic trek for both sheep and *pastor*. These green lanes through some of the remotest parts of the Iberian Peninsula represent a remarkable environmental heritage survival, in terms of their sheer scale as well as the wildlife reserves and corridors that they constitute: multiply 125,000 km of *cañadas reales* by their breadth – formerly up to 75 m over some major sections – and the answer is about 1% of the country's entire surface area. In Madrid, several main roads remain part of the *cañada* network, and symbolic sheep drives across the modern city are regularly organised to celebrate and defend this traditional transhumant practice. While the *cañadas reales* are known to date from medieval times, they may be much more ancient, possibly dating back to Neolithic times. The most important were established by royal decree, hence the term '*reales*' – though it seems highly unlikely that any members of Spanish royalty would have elected to pursue the lonely life of the shepherd, spending weeks or months away from the palace tending livestock along these remote, silent paths.

Transhumance was once the focus of a whole way of life, and over many generations this economic, environmental, social and cultural phenomenon has transformed landscapes and biodiversity across the Iberian Peninsula. Transhumance is the seasonal migration of livestock, and the people who tend them, typically between lowlands and adjacent mountains. Despite its low productivity, Bardenas Reales is a valuable resource for livestock farmers, and in mid-September each year El Paso near the small town of Carcastillo plays host to the dusty flocks at the end of their trek south along the Cañada Real de los Roncaleses from the lofty green Pyrenean pastures of the Roncal and Salazar valleys. Their annual entry to the winter grazing lands of the Bardenas is traditionally marked by the firing of a shot, after which they disperse among the hills of the hostile desert to glean whatever slim pickings they can. La Sanmiguelada is a festival held to mark the annual Día de Los Usos Tradicionales de la Trashumancia en Bardenas, featuring shearing demonstrations, and tastings of lamb, traditional *chistorra* pork sausages, wine: of course, and breadcrumbs. Yes... breadcrumbs. 'Shepherd's breadcrumbs' were produced by drying and breaking up stale bread to serve as sustenance for the arduous journey from the Pyrenees to Bardenas, and eaten with whatever came to hand including oil or butter, garlic, grapes, melon, paprika, tallow, chorizo, or pieces of ham. Possibly originating from North Africa, this unsophisticated but sustaining countryman's meal has now achieved celebrated status as a traditional dish. But while *migas cañadas* has survived into the twenty-first century, some of the *cañadas* have not, for although these traditional rights of way for sheep are legally protected from occupation and barring in perpetuity, the law has been much flouted across the country in recent decades.

Wandering along a dusty gravel track trying unsuccessfully to avoid the midday heat of late July, I struck up a conversation with José, perhaps the only other human for several kilometres on that day. Well, I say conversation: I had precious little Spanish and he no English, but somehow we managed to get on very well very quickly. Accompanying José were one dog and a lot of dusty and lean sheep. They were passing through; their journey had commenced in central Spain and would terminate in France. I was thrilled. Why was this solitary but friendly Spanish guy taking 200 sheep into France, I queried? He patiently explained that while sheep wool had little value in Spain at the time, it was prized in France, and in light of this obviously significant economic differential he was shepherding his animals 150 km across the border. The route necessarily involved traversing the Pyrenees, the second-highest mountain range in Europe, a journey he would accomplish with no equipment or baggage save a large coat, and very clearly sleeping al fresco at night throughout. It was worth it, he said gleefully, because in France he would make *mucho dinero!* (Twice, because I didn't understand him the first time!) After a fascinating if fleeting glimpse into the world of the itinerant Iberian shepherd, we parted as friends, and the last I saw of José he was striding purposefully north in a cloud of white dust surrounded by his temporary flock of inquisitive sheep.

José's dusty and lean sheep heading for France.

Numerous archaeological sites indicate that Bardenas Reales was settled in the Bronze and Iron Ages, and subsequently became the property of the royalty of Navarre, which is the derivation of the meaning of the word *reales*, meaning royal. The territory served as the king's hunting reserve and playground in much the same way that other parts of Europe did, with parallels in France, Germany, Italy and UK among others. After the Roman Empire, the territory was invaded by the Maghrebi conquerors of Iberia, when it is thought that Bardenas marked the northern frontier of their occupied lands. In recognition of the support given in the Reconquista by the men of the north, in 882 the monarchy of the kingdom of Navarre granted usage rights over the Bardenas, including rights to herbage and to build corrals (sheep pens) and cabins there. Usage rights were extended to the so-called highlanders of the Pyrenean pastures of the Roncal Valley and elsewhere, linked to the more southerly Bardenas by *cañadas reales* drovers' trackways. Subsequently, towns and communities surrounding today's park including Arguedas were granted some rights in recompense for military and other aid provided, and also to consolidate the reconquered towns and encourage the repopulation of the southern area of the kingdom of Navarre.

By the start of the eighteenth century, several towns and villages had established a variety of rights over the territory, for which payments were made to King Carlos II. Then, in 1705, 22 municipalities and the La Olivia monastery with defined privileges over the area paid 12,000 reals for the transfer of exclusive rights in perpetuity. The Bardenas developed an inclusive commons governance regime with open access to the area's resources for rights holders, the only barrier to entry being entrance fees for stock breeders and user charges for others. Livestock farming was the most significant land use in Bardenas Reales until the beginning of the twentieth century, with historical accounts putting sheep numbers at 300,000 head or more in past times. It's hard, lonely and low-paid work for the shepherds. Farmers have been growing cereals including barley in a few more fertile locations since the early twentieth century, although in such an environment crop yields will clearly be modest. Livestock uses have been giving ground to privileges in favour of arable agriculture, leading to a mosaic landscape of rain-fed cereal crop fields separated by unkempt patches of remnant natural vegetation, which nonetheless enable the maintenance of biological diversity.

To the north-west of the region, I happened upon what I sense was a demonstration of the capability of twenty-first-century agricultural techniques in the form of a large field of recumbent tomato plants. On first impressions, this meticulously irrigated attempt to make the desert bloom seemed pretty successful in conventional terms, with a fine-looking crop of scarlet tomatoes from a dry and dusty open field in prospect. However, knowing that the main driver of biodiversity loss in Europe is agricultural intensification – considered the major cause of the decline of farmland birds across the subcontinent – I couldn't help but conclude that producing a marginal contribution to the European salad mountain in this way at the expense of the scarce and threatened habitat that is the Bardenas seemed very short-sighted.

BARDENAS REALES HISTORY

This dry terrestrial biome was substantially more vegetated in former times than it is today. Present-day vegetation reflects not only the natural conditions but also centuries of human exploitation; some might term it a degraded landscape, with grazing by sheep and goats probably being the main contributor to deforestation over the long term. Historical records seem to confirm that around 1,000 years ago Bardenas was at least partly covered with woodland in a mosaic with treeless areas, though La Blanca may have had little or no oak and pine. The first records of logging here date back to the end of the eleventh century, uncontrolled and with catastrophic environmental impacts including an acceleration of erosion on the newly exposed soils. Wood was extracted for construction and firewood, large-scale clearance enabled agricultural expansion and 'excessive' levels of charcoal production combined with continued over-exploitative grazing and devastating fires to change the face of much of the territory. Nevertheless, although at the end of the sixteenth century Bardenas was still abundant in pine, wood exploitation for industrial uses continued until the end of the nineteenth century, and today only relatively few

Aleppo Pine is a Mediterranean native, thriving in full sun on dry soil, drought-resistant and with a beautiful rounded crown. (Christian Ferrer, CC BY-SA 3.0)

Kermes Oak is a slow-growing, dense, bushy, evergreen shrub or small tree resistant to grazing and drought and thriving in semi-arid habitats with dry soils. The scaly acorns are appreciated by cattle. (Javier Martín, Public Domain via Wikimedia Commons)

remnants of open Aleppo Pine forest, Kermes Oak ('scrub oak') and juniper break the horizon outside the treeless La Blanca.

The historic threats of overgrazing, desertification and the overuse of fire to renew pastures here continue to concern the *congozantes*, the 22 members of the common-pool resource institution charged since the early nineteenth century with guardianship of the territory. Yet as traditional means of land exploitation become increasingly uneconomic, land abandonment is also a twenty-first-century issue, threatening to trigger processes of natural succession. Grazing, and particularly sheep grazing, is essential for the long-term maintenance of this habitat, implemented at an intensity appropriate to promote biodiversity but avoiding overgrazing which will have a negative effect. Any sustained change towards a lower grazing intensity will result in scrub encroachment, changes in floristic composition, a reduction in biodiversity, and an increased risk of wildfires, resulting in the gradual substitution of the scarce European pseudo-steppe community. Maintaining traditional land management practices here is essential to maintain conservation values; at a global scale, many similar areas have been lost through conversion to cropland.

The human landscape is as distinctive in its way as the biophysical environment. No one lives here today. Indeed, to many, the prospect of living in this desolate European desert sprawling over an area larger than the Isle of Wight would seem hellish. A few abandoned small former houses excite some curiosity, and several ruined huts were perhaps summer quarters for shepherds, but quite where they obtained water for themselves and their flocks in this harshly evaporative landscape can only be guessed at. But another form of human occupation has entered the folklore around Bardenas Reales. Long considered a dangerous place, in the Middle Ages this otherwise inhospitable territory was notorious as a refuge for bandits, who were ready, willing and able to ambush unwitting travellers and rob locals too. While records are understandably scant, and accounts vary widely between sources, stories and legends of former banditry here are well known to this day. The former landscape of lush forests covering this location in the border regions of Aragon and Navarre was conducive to concealment, making the Bardenas ideal as a shelter for bandits, outlaws and fugitives. 'Going to the Bardena' was synonymous with fleeing justice, and regional through-routes to Pamplona, Zaragoza and Huesca would have provided a constant stream of booty for *bandoleros*. It seems that at the end of the wars against Castile and Aragon around the year 1200, highway robbers lived here as outlaws, most of them ex-soldiers marginalised in society and accustomed to prey and plunder. Things got so bad in this European outback that in 1204, 20 towns affected by this continuing scourge were forced to form an alliance to defend themselves. And they weren't messing about: several fortresses were built in Bardenas territory in the thirteenth century to serve as surveillance and control points, and specific local regulations mandated that wrongdoers caught red-handed would be dealt summary justice, being hanged on the spot with no recourse to conventional legal processes.

The most infamous and feared bandit of all was Sancho Rota, widely known as Sanchicorrota. In the fifteenth century, Sanchicorrota, a humble miller, was forced to flee his native town for killing a neighbour who worked for the king in a violent argument, and naturally he took refuge in the nearby wilderness. Over time, Bardenas Reales came to be regarded as Sanchicorrota's domain as he and his gang of about 30 ruthless men on horseback perpetrated ambushes, kidnapping, robbery and other crimes including looting and terrorising local towns. Like other fictionally celebrated outlaws elsewhere in the margins of western Europe including Robin Hood in England, Sanchicorrota is

Bardenas Reales refuge. (A1AA1A, CC-BY-SA-4.0)

said by some to have enjoyed a good reputation, the narrative being that he stole from the wealthy to give to the poor, and also that he was courteous towards those whom he robbed, provided that they were submissive and offered no resistance. Legend has it that Sanchicorrota's horse had its shoes fitted back to front to throw pursuers off his track, and another tale relates that he employed several associates to dig a cave in a remote Bardenas hill summit as a hideout – similar perhaps to those in Arguedas – but on completion promptly murdered his workforce to keep the place secret. A 425 m peak in the Bardenas is named Sanchicorrota – perhaps this was the location of his hideout. After endless provocation, in 1452 King Juan II assembled a reported 200 horsemen who surrounded the lair of the King of the Bardenas and massacred most of his supporters. Refusing to surrender, Sanchicorrota committed suicide with a dagger before he could be taken. To reinforce the message being sent to anyone else foolhardy enough to consider a career as a *bandolero*, his corpse was paraded around several local villages before being taken to Tudela, the regional capital, where it was hung on public gallows for the former bandit's flesh to be devoured by vultures.

But Sanchicorrota was only one of the thugs and lowlifes taking advantage of the opportunities for criminality offered in this European desert to make an easy living from others' misery. In 1120, the Bishop of Porto was obliged to disguise himself as a beggar in order to travel through the region without attracting the attention of murdering thugs. The bandit Moneos, said to have been a contemporary of Sanchicorrota, stole a shipment of fish but was traced because of the smell and captured. In 1590, Gaspar de Malla and his henchman Bustamante were running a gang of 50 brigands raiding various locations in Navarre, and despite the viceroy assembling a party of 300 men from Tudela and 150 from Ejea de los Caballeros, the perpetrators could not be found. In 1657, a gang of bandits robbed eleven muleteers transporting a cargo of valuable oil, silk, saffron and almonds in Bardenas de Caparroso. Forty residents of Arguedas and Valtierra went in search of the perpetrators but captured only one, who confessed that the gang had shared the loot equally as good friends. Stagecoaches were frequently robbed; the palace in Mélida was attacked and its owner, Doña Josefa Lapuerta, tortured and burned; and in 1532 ordinary citizens from Villafranca setting out to hunt and collect firewood in the Bardenas were advised to arm themselves against bandits beforehand. Despite the harsh penalties, banditry is said to have continued here into the nineteenth century, when the advent of new communications and the destruction of much of the remaining woodland through intensive logging finally helped to bring this brutal and criminal way of life to a close. However, the Bardenas remained a place of refuge well into the twentieth century, notably in 1936 during and after the military coup which sparked the Spanish Civil War, a time when political executions were taking place in nearby rural towns and Republicans seeking to escape fled to the relative safety of the desert.

MILITARY

In common with similar locations elsewhere in Europe, the face of this wilderness is disfigured by an extensive military zone. Smack in the middle of Bardenas is a huge fenced-off rectangular *zona militar*, enclosing a major practice-bombing range and airstrip used by the Spanish Air Force and others including the USA. Noisy planes of war fly in and out of what may be the largest military zone in western Europe, shattering the tranquillity of the three European Commission Habitats Directive priority habitats located here, although the military area has been excluded from the Biosphere Reserve designation. Stern-looking soldiers on patrol eye the casual adventurer with suspicion, and roads are closed at night, although that does not always deter those few determined to camp out seeking the full eerie, black-skies Bardenas night-time experience. Inevitably, the continuing military presence over about 5% of the area of this otherwise tranquil backwater brings risk: in 1980, the Spanish national daily *El País* reported a José Mª Aierdi Fernández de Barrena of the Basque Nationalist Party in the Navarre Parliament mourning the death of a shepherd from Arguedas killed in the crash of a Phantom fighter-bomber jet, a tragic event that in such a remote area must rank as unlikely as it was unfortunate. In 1979, a dropped bomb fell on a camping site in La Sotonera, Huesca, which is located 100 km from the designated impact zone! Happily, only dummy bombs are used on the Bardenas range. In another incident in 1982, a helicopter collided with the early seventeenth-century Yugo hermitage located on the hillside outside Arguedas. Perhaps it isn't surprising that local opposition to the continued military use of the site has been growing, though the national government avers there is no alternative, and the lease brings €14 million annually in a contract between the *congozantes* and the air force.

Back in Arguedas, the streets are still quietly sweltering. And the people are quietly leaving: the region is part of *España vacía*, 'Empty Spain', a term coined by the author Sergio del Molino. Tending sheep 365 days a year is not an attractive prospect for the youth of twenty-first-century Spain, and it is intriguing to speculate on the degree to which the relatively extreme climate prevailing here might also have contributed to individuals' decisions to outmigrate. The phenomenon of rural depopulation prevails over more than half of the country's surface area. In 2019 in the region of Galicia alone, 600 km to the north-west of Bardenas, there were an estimated 3,562 abandoned villages.[2] A lack of public services and opportunities in general prompt young women in particular to leave the countryside for the towns and cities, sparking the 'revolt of empty Spain' which has brought thousands of Spanish citizens onto the streets of Madrid to protest against depopulation and lack of investment in the regions. Though often invisible to the casual observer, rural depopulation is proceeding over much of the Mediterranean region, bringing socio-economic decline and large-scale land abandonment, though perversely there may be benefits for biodiversity.

POPULATING DESERTS

Bardenas Reales has seen several attempts to populate the dusty wilderness. In 1768, Jose Mariano Monroy proposed the creation of 18 new towns, and in 1772 another project was presented to found six new villages, though water supply and irrigation presented fundamental problems, and whether these projects were successful is unknown. More

---

2   Esther Costa (2019), 'How to repopulate rural Spain? Sell its villages', BBC Worklife: https://www.bbc.co.uk/worklife/article/20191121-can-tiny-abandoned-towns-put-galicia-on-the-map

recently, between 1945 and 1970 the state's Instituto Nacional de Colonización built around 300 new villages and towns, seeking to expand the national area under arable cultivation, boost farm production for domestic consumption and export purposes, and address rural depopulation then occurring as landless labourers, small tenants and sharecroppers headed for the cities in response to agricultural change. By 1973, half a million Spaniards had been resettled onto independent family farms in several clusters of new *pueblos* across rural Spain, a considerable national achievement by any standard. Where were they located? Yes, most were built in the *secano*, the drylands of Spain. In these remote and underpopulated wilds, the provision of irrigation was an essential precursor for the introduction of arable agriculture, entailing the construction of dams, reservoirs, canals and networks of distribution channels over a large area. Several such new villages were founded on the eastern margins of Bardenas in the mid-twentieth century. The intention was that the inhabitants of these modern rural utopias would work towards becoming owners of their small farms. To that end, the colonists were provided with a range of support including houses, land, tools and machinery, materials including seeds, education and advice, targets and standards for the production of specific crops, and an element of supervision. The *pueblos* were built with new transport and communications networks, health and education facilities, social and cultural provision, churches and shops – a long list that offers some indication of the level of public investment required to make the deserts of Europe bloom. Results seem to have been mixed, and in recent years employment in nearby towns has been becoming more attractive as agriculture has declined relative to manufacturing and service industries including tourism. Furthermore, few of the new *pueblos* have proved able to retain young adult population segments, suggesting that this ambitious rural colonisation programme seems unlikely to make a significant long-term impact upon the Empty Spain phenomenon.

Several national governments have attempted to populate, exploit and civilise the world's deserts by attracting settlers there. The Montana prairie and badland drylands of the north-western USA were promoted as ripe for settlement to restless, independent-minded city-dwellers through pamphlets and newspaper adverts promulgated across the USA and several European countries including Germany and Scandinavia. As new railroads were being extended towards the west coast during the 1880s, an infant city was established every dozen miles or so. Tens of thousands were moving to arid Montana each year by the second decade of the twentieth century, lured by gaining title to land which was virtually free, and the prospect of building their own homestead on the largely empty plains of the Big Sky Country. The US administration backed the homesteading of the treeless, sparsely vegetated dry west to relieve population pressures in the overcrowded cities, and to boost national production of grain and cattle on farms small in scale compared to the old-style ranchers' operations. But the inhospitable environment and unpredictable climate combined with inexperience, poor soil, no irrigation, sun-baked crops, starving cattle, freezing winters and a series of dry years around 1920 saw the land being abandoned in a mass exodus, with many heartbroken homesteaders heading back to the cities, leaving behind deserted shacks and untilled fields. As a result, many of the state banks went bankrupt as loans could not be repaid.

A European desert par excellence, Bardenas Reales is a vast, parched, empty territory with a true desert climate where temperatures at times rival those of the Sahara at its most extreme. With the character and atmosphere of a stark, desertic wilderness at the ends of the earth, here the dominance of natural processes and the physical environment – both mineral and climatic – is unequivocal. It is hard not to conclude that wildlife is much more adapted to living here than puny humans. Yet this Biosphere Reserve and Natural Park is simultaneously a cultural landscape, managed and exploited for millennia like many other European deserts. Despite the depredations of the twenty-first century nibbling at its margins, much of the Bardenas has survived and will continue to do so. Among the landscape drama, huge skies and spectacular wildlife, the grubby but proud mystery and sheer naked elemental majesty of this peerless European environmental phenomenon will stay with you forever.

Chapter Three

THE NETHERLANDS: DRY WILDERNESS IN A WATERY COUNTRY

It may seem a little incongruous to find a chapter focused on the Kingdom of the Netherlands in a book on European deserts. This small low-lying country on the far western seaboard of Eurasia is better known for water than it is for desert. Much of the territory is dominated by a river delta system and water is a major feature of the national landscape – nearly 20%, to be precise. Dutch history and culture are closely linked to surface water and its management: polders (land reclaimed from the sea), dykes (walls constructed to keep sea and rivers out of polders), canals and the windmills once necessary for pumping water away before diesel and then electricity. A full third of the country lies below sea level, and with a long history of flooding the Dutch have spent centuries keeping the sea out while reclaiming land, about one-fifth of the nation having formerly been sea and lakes. In a country that mostly comprises flat, low-lying plains – hence the Nether-lands, in the Low Countries – and with plenty of fertile soil, the Dutch national phenomenon of high levels of agricultural production has made the diminutive kingdom the second-largest exporter of agricultural goods after the USA. Almost every last square metre of this impeccably civilised and quietly wealthy place is deemed valuable and meticulously managed.

Yet amidst all the carefully planned agricultural landscapes, fastidiously laid-out settlements and ubiquitous waterbodies are hidden some of the most outstanding instances of the European desert phenomenon. The deserts of the Netherlands are unrivalled in north-west Europe for the spectacle of desertic wilderness they offer, and also their current (and former) extent. Many exhibit classic desert-like features including large expanses of open sand, rolling dunes and the potential to drift beyond their boundaries, threatening to swamp farmland and forest, villages and roads under blown sand. How did the Dutch deserts survive in this minutely managed small country? How did they develop as they have? How has their existence impacted upon their surroundings? How and why are some now actually *expanding* in area?

As a predominantly sandy place, the geology of the Netherlands is ripe for European deserts to flourish. Like water, sand is a major feature of the landscape, although this fact is not immediately obvious to the casual observer. More than half the country is covered by sand, in the form of vast sheets frequently with dunes, and sand fringes most of the coast too, from the Danish to the Belgian border. Numerous European deserts are located inland. Formerly much more extensive than today, they are a ubiquitous if not well-known element of the Netherlands' national landscape. The Dutch people have grown to treasure them, offering as they do a glimpse of the planet in the raw, scarce havens for wildlife, a reminder of historic environments and societies and a flavour of the planet's subtropical sandy deserts. During the course of several research expeditions, I headed for one of the best examples.

Apeldoorn is a respectable medium-sized country town in the central Netherlands. Striding casually into the municipal tourist information centre there one late May, I must have been easily recognisable as a tourist – indeed, the two staff there instinctively addressed me in English. The range of visitor literature made clear that the town is not over-endowed with tourist attractions, apart perhaps from the royal family's former summer residence, the seventeenth-century baroque Paleis Het Loo, an austere royal palace set within finely manicured formal gardens on the edge of town. The friendly and attentive tourist information staff furnished me with a map of the district, but looked a little askance as I patiently made clear that I hadn't travelled all the way from England to view regular tourist attractions. I enquired about a highly irregular one: Kootwijkerzand. They both stepped back for a moment, obviously puzzled. Surely this English guy would not have travelled so far to visit the local *heide* (heath), that dusty wasteland 15 km out west of town? Gently I convinced them that, yes, I really was here to see Kootwijkerzand. Which bus should I take, and how long was the journey? Now recognising my needs, the friendly staff addressed my queries. I thanked them and bid them '*Afscheid*' (farewell) before heading for the bus station.

Alighting from the Amersfoort bus and heading south off the road, I plunged into dense conifer woodland. Very quickly the familiar landscape of Europe's most densely populated major country faded as the path threaded its way through plantation Scots Pine. Breaking into a clearing, I found Kootwijk village was tiny with several fields devoted to 'horsiculture' before once again heading into woodland. As I crossed Duinweg (Duneway), the daylight dimmed under the forest canopy once again, and I innocently pressed ahead.

KOOTWIJKERZAND

Nothing had prepared me for my first sight of Kootwijkerzand. As I burst out from the trees into the mid-afternoon sunlight, an intensely bright vista of sand drifts opened up before me. Sweeping expanses of pure sand ran to the far horizon. Bare, bare sand, interrupted only occasionally by isolated dunes, the odd gnarled, superannuated Scots Pine, a few scraggy bushes, patches of purple heather and dribs and drabs of desiccated grasses seemingly huddled together for survival. Now *this* was a desert, at 700 ha western Europe's largest active inland drift sand landscape. It really felt like the naked planet. Something like the Sahara.

Gazing out over Kootwijkerzand, the dominant prospect was of flat and gently rolling drift sand plains stretching into the distance. At random intervals were scattered low dunes, with bare sand on their windward sides and grass clinging to the leeward,

Netherlands desert at De Hoge Veluwe National Park near Arnhem.

sheltering from the wind and constant abrasion from mineral sand. Here and there were much larger plateau dunes, steep-sided and rising abruptly from this largely horizontal landscape. Sandy and with a roughly flat top, these 'blown over' buttes are founded on a buried impermeable layer which forms a perched water table here in the dry desert. A handful of wind-battered bushes and stunted trees clung onto these so-called forts. Nearby were 'blown-out' areas, large flat or concave expanses looking like smooth yellow amphitheatres where sand is continuously removed by deflation – wind blow – the most spectacular features in this distinctly unspectacular landscape. Wandering further into this wide-screen Dutch sandpit, I could feel how the interplay of wind and sand dominate. Simply walking through these soft sand seas was draining as my feet sank through the surface at every step, rapidly sapping energy and the will to continue. My bare legs were being subtly but distinctly eroded by the abrasive sand particles continually entrained over the land surface by the wind.

Dry dry dry. There was absolutely no water, or water features. How do the few higher plant species able to survive here manage to obtain water reliably on these bitingly arid and low-nutrient sands? Sand is more normally associated with low altitudes but here a huge pile of the stuff rises to an uncharacteristic Dutch altitude of 110 m, the highest point in the central Netherlands. With such a dominance of sand, Kootwijkerzand felt more like a coastal than a terrestrial inland environment. But of course this was a misperception, for the tract lies inland, 100 km from the nearest sea coast!

Kootwijkerzand lay deserted in mid-afternoon. There were no people. And there was almost no sign of people either: no structures, no artefacts, no walls, no hedges, no ditches, no footprints (for they disappear almost as fast as they are made under the continual assault of wind-plucked sand), not even any litter. A weird and somehow slightly suffocating atmosphere hung over the place. It was practically silent too, as are all the European deserts. These are quiet soundscapes. Sand on such an extensive scale attenuates all sound. There are no echoes to be heard on the deserts of Europe. No tree leaves rustling, no streams tinkling, no reeds shushing, what wildlife there is produces little or no noise, and there are usually no anthropogenic sounds either because most sites are remote from human activities. You might term it desolate; I did not. Somehow the elements seem more pronounced here: the weather seems more intense than elsewhere. Though not yet summer, mid-afternoon temperatures on the sun-drenched sands seemed unexpectedly high, but was the temperature really higher here than in the nearby surroundings; does the sun actually shine more brightly here in the Dutch deserts? It almost seemed so. When it rains, too, it seems somehow more spectacular than elsewhere, with the sandy ground surface dancing to the beat of the raindrops, and there is absolutely no shelter for the ill-prepared. Was the wind really more persistent here than elsewhere? There did seem to be a near-constant breeze. And when the wind blows hard, sandstorms are the result...

Kootwijkerzand panorama. (Frans Berkelaar 2021)

Bare sand and dunes in the Hoge Veluwe at Schaarsbergen. (Henk Monster, CC BY 3.0)

Despite the distant pine fringe around Kootwijkerzand, it's easy to lose one's usual sense of scale and perspective here. There are few reference points, few landscape features of any kind in fact, and the white-yellow sandy vistas recede and blur into the distance, mysterious in their mute weirdness. Unlike elsewhere in civilised Europe, with its established and familiar agricultural landscapes, one soon appreciates that this is a landscape of impermanence. Little is fixed, little is forever. Kootwijkerzand vistas evolve constantly over time as the dusty elements of the landscape are shuffled and reshuffled by the wind. It's unsettling like nowhere else. The very idea of drift sand is anathema: not solid, not secure, not enduring, not risk-free, and it fails to offer any sort of dependable natural foundation. Almost everything here seems unbounded, unfixed, unfinished, unsettled, unresolved, even uneasy, uncomfortable. Inland sand dune landscapes feel almost surreal, and certainly not at all what you might expect in the Netherlands. And when in winter the snow arrives, the vistas change eerily from Sahara to Siberia.

INLAND SAND DRIFTS

The Dutch drift sand landscapes comprise some of the least-known yet most valuable landscapes in the country and indeed western Europe. Numerous instances are designated as nature reserves and national parks. Characterised by large expanses of barren shifting sand set within a mosaic of moss, lichen and pioneering grasses, in twenty-first century Europe *active* drift sands on such a large scale are confined almost exclusively to the

Netherlands. Only in a few European locations can a succession of habitat types grading from bare ground to established woodland be observed. Here is a unique ecosystem characterised by a floral and faunal composition that is species-poor but nevertheless highly distinctive. Many species are specialists adapted to the exceptional habitat conditions, for drift sand presents a challenging prospect for many organisms due to the arid, acidic and nutrient-poor substrate, continuously changing environmental conditions and widely fluctuating daily temperature ranges. Pioneer vegetation faces challenges taking root, and sustains mechanical abrasion when blasted with sand particles; algal mats are continually buried. With a low water retention capacity, extreme variations in soil moisture content, and the water table lying anything from 20 to 50 m below the surface, the hungry sands suck rain droplets downwards and away from the root zone in the absence of a more fully developed soil structure. Some plants are dependent upon precipitation for water and atmospheric deposition for the nutrients they need. My brief but inspiring time here led me to understand why all over Europe these areas were termed wastes. Cultivating conventional crops on drift sand is impracticable. Kootwijkerzand and others really are 'bad lands' in conventional terms of agricultural productivity, dramatically different from the fertile fields of the surrounding regions. Had these tracts of land not been so marginal in farming terms, they would have been 'improved' – converted to agricultural use – long ago.

Harebell tends to favour dry, open ground.

Geology is the basis of the existence of the Dutch deserts. Kootwijkerzand is located on the Veluwe sand massif, an extended ridge of hills and plateaus in east-central Netherlands. Here set among dense plantation forests are scattered a series of large drift sand expanses over a region 60 km north to south. Glaciers and their meltwaters brought large quantities of sand and gravel from the north, and this accumulated debris was subsequently deposited in the form of hilly ridges. The Veluwe region comprises a complex of such 'terminal moraines' dating from the Saalian glacial stage of 400,000 to 130,000 years ago. The sand was predominantly composed of quartz particles, the second most abundant mineral in the Earth's crust and the hardest everyday mineral, resistant to both chemical and physical weathering. In cold polar desert-like conditions when northern Europe was ice-free, these sediments were subsequently reworked over millennia by intense wind action. Sand grains were whipped up and slammed together repeatedly in weathering processes to produce drift sand in the form of very fine spherical mineral particles with opaque surfaces. Vast quantities of these inert aeolian (wind-borne) sands were redeposited over northern and north-west Europe – blanketing large areas and burying earlier river landscapes under a flat to gently undulating relief – in an unconsolidated form ready to be cast aloft again, reactivated in raw, windy, exposed environments.

Pathside wind-erosion feature at De Hoge Veluwe National Park near Arnhem.

A prospect of bare sand in De Hoge Veluwe National Park.

Active drift sand expanses are the source areas supplying the windblown sand essential for the continuous regeneration of this distinctive landscape. They can readily be discerned on aerial photographs, frequently appearing like Kootwijkerzand as elongated oval-shaped cells of 1.5–6 km in length oriented in the direction of the prevailing south-westerly wind. Active sand drifting recurs over unvegetated areas during intense storm events. Each cell exhibits a three zone structure: sand is lofted into the air from an open deflation plain – or blow-out zone – of pure wind-whipped sand to the south-west, where the ground surface is physically lowered by aeolian action, before being entrained across a transportation zone and finally deposited in an accumulation zone to the north-east. Here, large quantities of sand are deposited in the surrounding vegetation, and dune formation may occur.

The drift sand areas were gradually colonised with trees over millennia until the Neolithic Age about 5,000 years ago, when the earliest farmers settled on the higher parts of the Netherlands resulting in a relatively dense pattern of settlement on the Dutch sand plateaus. The sandy soils were well drained and easy to till with simple implements, and over time the woodland was gradually destroyed and replaced by open heathland. While there is evidence for local sand drifts from Neolithic to Roman times, the start of extensive sand drifting here and elsewhere in north-western Europe is thought to be correlated with the rapid expansion of agriculture and over-exploitative grazing, burning and nutrient-depleting practices on heathland where farmers held the land in common for sheep grazing from about AD 950. Much of the Veluwe developed into a vast expanse of drifting sand, and after about 1150 the area of sand expanded significantly as the pace of deforestation accelerated. Periods of severe drought, strong winds, a dwindling water table, and disturbance wrought by roads, paths and even sheep tracks could generate sand drifts.

The resources of the Dutch deserts were limited but nonetheless worth exploiting. Grazing continued throughout subsequent centuries, and combined with burning practices led to the landscape retaining its open character. Extensive bare sand surfaces vulnerable to wind erosion were maintained by growing anthropogenic pressure in combination with natural causes, notably climate changes influencing the frequency of storms. During dry and warm climatic periods, the winds were free to play with the sands producing extensive drifts. Landscape historians consider that the main expansion of drift sands probably dates from the thirteenth or fourteenth century, when Kootwijkerzand turned into a real desert and sand drifts started to threaten the region. Areas of bare drift sand expanded during extreme erosion events as the deposition of large quantities of infertile sand meant that, once covered, surrounding areas of vegetation would wither and could not regenerate. Worse, unpredictable sand floods swamped fields and houses. One village disappeared under 2.5 m of sand. Crops failed. Water wells dried up. It was a precarious existence. Farms were abandoned and villages deserted. Gradually, whole areas were abandoned to natural forces. Over subsequent centuries, drift sand areas continued to increase in extent until the latter half of the nineteenth century, when the Veluwe region had developed into one of the largest deserts in Europe at about 15,000 ha.

From the fourteenth century, the fight against the deserts became a major priority, occupying the Dutch authorities here and elsewhere until the twentieth. Attempts were made to control wind erosion and sand drifting on the Veluwe with the appointment of a local government fire officer (*brandmeester*) around 1560, whose duties were to control heath and forest fires and regulate grazing. In some locations, firing the commons in the traditional manner became punishable by the death penalty. Around 1650 the office of Zandgraaf van Veluwen was created by local government, and a sand reeve and assistants (the 'Sand Board') appointed with the specific task of controlling the sand drifts, but they had only limited powers to intervene. They could order farmers to take land management measures to control the transgressive deserts, including planting hedges and erecting withy fences to protect against wind and sand, and covering bare sand expanses with heather sods. But there was no decrease in the area of sand: instead, the area increased, because of a shortage of money to implement control measures – most was said to have been spent on the salaries of the officers – and partly because of a lack of cooperation from land users. Conservation measures were hampered by high fines for those misusing the land, which impoverished land users simply failed to pay. Conservation efforts in the more remote areas – those far from settlements – were often only half-hearted, and led to disputes about who should pay for the work. Population growth and resource exhaustion continued to increase pressure upon the land. The Dutch deserts expanded to such an extent that in 1696, Edward Southwell, an Anglo-Irish lawyer and politician, observed that the Veluwe wilderness was 'One of the best hunting Countrys in ye World but good for Little else'.[1]

DESERT INHABITANTS

The post-medieval countryside was marginal in social terms, and large areas of heath remained uninhabited until the mid-nineteenth century. Outsiders regarded the region as barren, remote, hostile and unsafe, peopled only by farmers and shepherds. Life on the Veluwe was so challenging that for centuries most rural people avoided living there, preferring instead to dwell in the villages and estates of the region. Only the poorest people, such as broom-makers, had their humble cottages and huts at its edge. The role of the heaths included functioning as a buffer or fall-back resource in times of economic downturns and scarcity. Everywhere travelling peoples of all kinds are attracted to such areas, and the first records of itinerants date from the early fifteenth century. By the seventeenth century, difficult economic conditions in the Netherlands led many people to turn to begging, and when their numbers increased significantly in the eighteenth century vagrants and landless paupers were increasingly considered a problem. Some settled communities appointed an *armenjager* (Dutch: pauper hunter), whose role was to expel the destitute poor from common land or that belonging to a settlement. Beggars and vagabonds were persecuted and driven away – often with some zeal – using dogs, sabres, pistols, and muskets as necessary. The unfortunate pauper was usually expelled from the municipality by being forced across the border into the neighbouring district, thus making them somebody else's problem.

However, from the seventeenth century social attitudes to the countryside were changing. Elite groups searching for idyllic rural retreats began building castles and establishing estates, country houses, landscaped gardens and water features on the margins of the area where water was more available. By the nineteenth century, writers and painters too were discovering the rugged beauty of the Dutch wildernesses, and in their wake came tourists staying in new luxury villas and hotels. Sanatoria were built to treat tuberculosis patients, along with convalescence and retirement homes.

At the other end of the social scale, in the early nineteenth century a benevolent society aiming to improve the lot of urban vagrants and beggars sought to create several new agrarian 'pauper colonies'. In a large-scale poverty relief social experiment, poor people were relocated from towns and cities to the countryside where they were to learn to cultivate cash crops to be sold into the market. The north-eastern Dutch province of Drenthe was selected as the location for several new villages, a remote, sparsely populated and poverty-stricken region with extensive areas of uncultivated sandy heath and wetland which the colony projects would 'reclaim'. Settlements were created from 1818 at Frederiksoord, Willemsoord and Wilhelminaoord, where the new ruralites were

---

[1] Neefjes, J. (2018) Landschapsbiografie van de Veluwe Historisch-landschappelijke karakteristieken en hun Ontstaan. Amersfoort: Rijksdienst voor het Cultureel Erfgoed.

put to work reclaiming large areas of wasteland. The pioneers were given modest houses with a little land and some household goods and tools, the intention being that several years of hard work in the countryside would turn them into industrious workers able then to leave the colonies with some savings and an improved mentality. Estimates indicate that an extraordinary 800,000 ha of desolate and unproductive land around the colonies was transformed into agricultural land – every hectare reclaimed through back-breaking toil – with new villages laid out with neat buildings set alongside straight avenues.

Suitable colonists willing and able to toil as capable and self-disciplined sharecroppers proved hard to attract, however. There seemed little enthusiasm for what were regarded as covert labour colonies on the margins of Dutch society in isolated Drenthe. As some of the poor refused to work, intensive control and even coercion was employed including a strictly regulated regime of meals, work and rest which included only limited leisure time. Few migrated to Drenthe voluntarily: over each of the years from 1830 to 1860 an average of just twenty-two families arrived of their own volition. The colonies were supposed to be financially self-sustaining within sixteen years of their inception, but harvests were disappointing because the soil was not suitable for growing crops without the addition of manure sourced from outside at additional expense, and management lacked the necessary agricultural experience. In 1859, the near-bankrupt colonies were taken over by the state, after which thousands of families and individuals seen as undesirable elements of society were deported there by urban authorities.

Ommerschans and Veenhuizen, two villages established later in the 'Dutch Siberia', became unfree penal colonies, imprisoning paupers whose crimes were often more idleness, poverty, unemployment and drunkenness than lawbreaking alongside convicted criminals. Inmates were kept under permanent guard and forced to work on the wastes – a substantial area of wetland of 4 × 2.5 km was reclaimed at Ommerschans, home to 1,308 people in 1840 – and church attendance on Sunday was compulsory. Some slept locked into stacked-up horizontal metal cages more reminiscent of a battery chicken farm. Ordinary Dutch people were discouraged from entering these villages. Over the period 1818–1921, about 80,000 people moved to the countryside in Drenthe, and at their peak in the mid-nineteenth century over 11,000 people lived in colonies in the Netherlands. Today, most are regular communities. Veenhuizen village, one of these symbols of historic grinding poverty and shame, is now preserved as a visitor attraction and has gained UNESCO World Heritage List status; the village still has state prisons. Quite how benevolent the Drenthe colonies were is a moot point: were they really 'paupers' paradises' by the standards of the time? These projects have been likened to the colonies exploited by the Netherlands in the former Dutch East Indies that subsequently became Indonesia in 1949. The new villages established in the Dutch wastelands of Drenthe have been characterised as having been dependent upon the exploitation of labour, and through processes of internal colonialism that mirrored those in the Netherlands' international colonies.

In 1939, another development was proposed by the Dutch government seeking to intern waves of Jewish refugees fleeing from the Nazi regime over the border in Germany. A need was identified for a central facility to intern and process the new arrivals. Where might such a sensitive facility best be located? Why, on the Dutch heaths! Proposals were made to construct a reception centre on a sandy Veluwe plain near the village of Elspeet. However, local residents protested, as did Queen Wilhelmina who objected to its proximity to Paleis Het Loo 14 km distant. The internment centre was finally sited in the remote province of Drenthe, about 10 km from the village of Westerbork and 40 km east of the pauper colonies. Few of the Dutch people knew the place existed then, and few do now. Shortly after construction, Camp Westerbork was repurposed as a transit camp, through which up to 100,000 Jews and others – including Anne Frank – passed to destinations in Nazi-occupied Poland. Here is another instance of how unwelcome land uses have often been exiled to the deserts of Europe.

BIRDLIFE

Notable fauna of the Netherlands deserts include heath habitat specialist the European Nightjar, a fascinating bird that is both crepuscular – active at dusk and dawn – and nocturnal. With pointed wings and a long tail, their shape is similar to a Kestrel or Cuckoo. Across much of its range, the Nightjar's breeding distribution is closely associated with conifer plantations with sparse tree cover adjacent to heathland. During the day this medium-sized bird roosts mostly on the ground, remarkably well camouflaged in its mottled grey-brown plumage, until dusk when the bird takes up a perch in the trees. One

European Nightjar perched motionless on a branch. (Imran Shah, CC BY-SA 2.0)

of the few insect-eating birds that hunt at night, the Nightjar expertly hawks flying insects including moths, beetles and flies over open heath habitat. Goatsucker, this enigmatic bird's traditional folk name, derives from an old myth that nightjars steal milk from goats at night. Their flight is powerful, agile and noiseless; enormous eyes adapted for night vision offer a large field of view to detect prey; and a wide beak gape scoops up insects with precision. During the spring mating season, the male occupies a territory marked out by its call, an unmistakable rising and falling 'churr'. Sometimes sustained for 10 minutes or longer, this eerie and relatively loud nighttime trill characterises these environments. Airborne wing-clapping is the bird's way of attracting a match, a curious trait also employed to chase away intruders which has been observed directed at deer – and also inquisitive humans, including the author. A summer migrant visitor from May, by mid-August the European Nightjar is heading south on an epic journey to its wintering grounds, scrub-dominated grasslands in sub-equatorial Africa. While there has been a significant increase in the Netherlands breeding population, the Nightjar remains a protected species whose main threats are lamentably familiar: degradation, fragmentation and loss of their favoured open heath and bare sand habitat; disturbance in general; and reduction of prey abundance through pesticide use.

The fate of another bird with similar habitat preferences, the Tawny Pipit, provides a valuable insight. Relatively few fauna species are capable of flourishing in the harsh conditions of the Dutch deserts, but sparse grassland and open dry ground is the favoured breeding habitat of this migratory bird. Though situated at the north-western extreme of

The instantly recognisable Silver Birch is a ubiquitous tree on heathland, supporting over 300 species of insect and attracting birds such as Greenfinches and Siskins for the seeds. (Flowerikka B on Flickr)

A lowland species, the habitat of the Sand Lizard is a combination of open land, dunes and vegetated patches. It spends most of the time basking, foraging, or under vegetation, and uses warm sand to thermoregulate itself and incubate its eggs. (George Chernilevsky, Public Domain via Wikimedia Commons)

the Tawny Pipit's Eurasian breeding distribution – primarily the semi-deserts of western and central Asia – Kootwijkerzand was formerly a core area for the species, with the large areas of open sandy habitat essential for its survival sustaining viable populations. However, the bird's steep decline and disappearance both here and elsewhere in the Netherlands in the twenty-first century has been attributed to habitat change and loss – notably through afforestation – along with disturbance caused by recreational use.

PLAGGEN

The stunning drift sand landscapes of today's Dutch deserts were fundamentally transformed by humans over recent centuries. The ingenious Dutch employed a novel though little-known system to fully exploit their sandy wastelands, under which they assumed a critical though slightly improbable role in supporting the fertility of nearby arable fields. A key phenomenon characterising the deserts of the Netherlands and others in the Low Countries and north-west Germany was the historic practice of 'plaggen', a system for transferring plant nutrients from wastes like Kootwijkerzand to arable fields. In the agrarian age, many Dutch European deserts were the focus of an extraordinarily invasive and labour-intensive agricultural system with the goal of improving the productivity of not-very-good local land by systematically and repeatedly plundering the very poorest local land. The plaggen system was practised for centuries over a huge swath of north-west Europe, yet ceased a century ago and today is barely remembered outside scholarly circles. The sandy Netherlands deserts lacked the fertility necessary to grow rye continuously. Just one year of rye cultivation on such poor soil would necessitate a subsequent fallow – uncultivated – period of up to nine years before the cereal might successfully be cultivated once again. How might the barely fertile deserts have helped to supplement arable agricultural production in the era before industrially produced fertiliser?

The plaggen system represented a level of sophistication in the vital process of maintaining and enhancing arable yields on poor soils. The Dutch word *plaggen* refers to the scraping or excavation of earthen sods from the ground. Sods of turf were stripped wholesale from the deserts and translocated to fertilise recipient fields which were mostly situated around the margins of the Veluwe. Extracted from the uppermost humus-rich layers of soil, the sods were formed of grass, herb and heather roots along with attached soil materials. Plaggen sods in the Netherlands were 25–30 cm square and 3 cm in depth, cut manually from the land with special tools in a labour-intensive process. They were the agents of a nutrient-transfer process translocating fertility to the permanent arable fields. Before application to the fields, the sods were used as bedding in cowsheds and sheepfolds, where they absorbed animal urine and excrement thereby boosting their limited fertility. Often composted for weeks or months afterwards, the plaggen material – by now valuable organic mineral manure – was spread over the recipient arable fields where it decomposed and humified, producing a soil with improved properties and conditions for plant growth. Repeatedly adding plaggen material in this manner – a process known as turf manuring – was considered more effective than fertilising with manure alone, producing soils richer in humus and with higher nutrient availability which in turn boosted cereal yields.

Over the centuries, repeated fertilisation by the plaggen method had profound effects. The depth of soil in the recipient arable fields might increase by one-third to one-half, and the land surface might be raised in height by 1.0–1.6 m over time. Also important was an accompanying doubling or even tripling of the soil water retention capacity. It was a big job. One estimate from research on the nearby German deserts suggests that for 1 ha of arable land to be raised in height by 1 m required 13,000 tons of plaggen soil to be sourced, transported and applied. Historic photographic images depict hard-working labourers digging sods of turf from the ground by hand amidst vast, flat, bleak, featureless landscapes. Today, plaggen fields are still clearly identifiable in the rural landscape, one impact of their presence being that some historic farmhouses are set at a level appreciably lower than the elevated fields surrounding them. Trial pits dug down today into former plaggen-treated fields growing arable crops reveal sandy former topsoil at the base, above which lies a very distinct boundary overlain with a substantial depth of thick black plaggen-derived soil which had been shovelled on top in former times. These are termed anthrosols, anthropogenic soils formed by historic land management practices which resulted in profound changes in soil properties.

The brawny Dutch peasant farmers repeatedly exploited the common wastelands for their meagre fertility over several centuries. This was the best workable method of maintaining the fertility of the permanent arable fields of the Dutch desert regions in the agrarian era. The large scale at which plaggen stripping on the sandy lands was conducted is affirmed by estimates that each hectare of intensive rye cultivation required an area of – variously – four, five, ten and even forty hectares of corresponding donor wasteland. Everything was accomplished by manual labour, of course, and the sheer enormity of the scale of human effort expended by skilled and hardy workers seeking to eke a livelihood in an unpromising environment can only be marvelled at. On larger farms, one worker's duties might have been solely dedicated to plaggen tasks. The only non-human powered assistance was from the horses and cattle employed to draw the carts transporting the plaggen. While there are few estimates of the historic extent of plaggen, the system was widely employed throughout the Low Countries and north-west Germany and may have prevailed over 10% of the Netherlands.

The plaggen system clearly worked sufficiently well in sustaining arable soil fertility to make reliable rye cultivation possible. Fields fertilised in this way might typically produce two to three times more rye than without, and remained productive for 20 or exceptionally 40 years without the need for the crop rotation or fallowing employed elsewhere. These soils retain improved properties and conditions for plant growth to such an extent that, even today, plaggic anthrosol fields are valued more highly at auction for certain specialised cultivation purposes, including tree nurseries. So, the economic impacts of centuries of hard manual labour by agrarian-age peasants in the Dutch wastelands continue in the twenty-first century. But what impact did plaggen have upon the Dutch deserts? Tracts of common land were stripped repeatedly over decades or centuries in a cycle of overexploitation of their natural systems, repeatedly denuding the scant nutrients accumulated there. This sustained nutrient-stripping and transfer process

The heideplagger – a labourer cutting turf sods in Hoven, Denmark. (Museum Midtjylland, Herning, Denmark)

arrested natural ecological succession over the drift sand areas, maintaining them as open and unvegetated. Plaggen perpetuated ecosystems of a low nutrient status, and also the shifting drift sand environments. The practice of plaggen was profound in its impact, scale and longevity; it had a fundamental impact on the ecology and landscape of sites like Kootwijkerzand; and probably too a substantial cumulative impact across the whole Veluwe region. Kootwijkerzand and similar tracts served as the equivalent of a contemporary agrochemical nitrogen fertiliser factory. The rate of extraction increased over the centuries, as the 'sod economy' intensified and became increasingly commercialised, resulting in a gradual decline of the quality of harvested plaggen. At the close of the heath farming era, as much as two-thirds of the drift sand areas may have been devastated by sod extraction, which would have produced the open sand landscapes we see today but on a vastly greater scale.

Over-exploitative land use had led to a process of soil degeneration, resulting in landscapes which have been described as 'exhausted'. Continually denuded of their protective vegetative cover, the fine sand was exposed to wind erosion and frequently rendered unsuitable even for sheep pasturing. Had depleted drift sand areas been allowed to recover over a fallow period after plaggen harvesting, they might have regenerated over a period of 20–40 years, but it seems that such a judicious approach was only rarely adopted. Overall, *plaggenwirtschaft* created and maintained vast bare surfaces on the Dutch drift sand deserts which over time became large areas of shifting sand, especially when combined with heather-burning and often uncontrolled grazing. Plaggen may have represented a 'robbing Peter to pay Paul' endeavour in some senses, but by continually retarding ecological succession in such an intensive manner the practice undoubtedly maintained the distinctive landscape character, biodiversity and perhaps the very existence of the Dutch deserts, which may otherwise have been lost under naturally regenerating trees a long time ago.

PLAGGENHUTS

Historically, living on the Veluwe was challenging, though housebuilding in a novel vernacular style flourished here. For centuries, most rural people strove not to live on the heath, preferring instead villages and estates. Only the poorest inhabitants, such as broom-makers, had their humble cottages and huts at its edge. Outsiders regarded the region as barren, remote, hostile and unsafe, peopled only by farmers and shepherds. The living conditions of the impoverished Dutch heath-dwellers were wretched. Some settled on drift sands remote from established Veluwe villages, where they endeavoured to cultivate vegetable gardens and arable fields. In need of accommodation, they built their own houses. With little money or conventional building materials, the poorest workers made do with whatever they could find on the heath to construct a dwelling known as a plaggenhut. This was a one-roomed single-storey cabin home inhabited by the poorest workers in both sandy and peatland areas of the Netherlands and elsewhere in Europe. Very different from conventional houses, these huts were constructed in a manner which dated back centuries, perhaps even to the Bronze and Iron Ages, and used the turf sods employed in the plaggen process for translocating nutrients to arable fields. Capable of

A derelict plaggenhut near Otterlo on the Veluwe sand massif. (Rijksdienst voor het Cultureel Erfgoed)

being built by the average individual and quickly, the traditional plaggenhut dwelling has been described simply as a dugout – a large hole with a roof over the top. An oblong pit of about 4 m by 6 to 8 m was dug to a depth of about 50 cm and the floor formed of tamped-down sand, loam or pebbles. Typically starting at ground level, the roof was constructed as a basic wooden frame structure of tree branches and slats. Sods of heather extracted from the surrounding dry land were placed on top in the manner of roof tiles. Walls were built by the simple stacking of plaggen sods. A wooden door would be fitted at one end, and windows too if funds were available, though these were typically fixed and could not be opened. Inside, the walls may have been finished with newspaper or wallpaper. Hut villages arose on the Veluwe and in other regions, their appearance blending harmoniously with the surrounding landscape. Often constructed in clusters to promote social interaction and mutual security, some of these slum settlements accommodated as many as several hundred inhabitants.

Life in the plaggenhut hovels was cramped, insanitary and unhealthy. Entire families of six to nine people lived together. The house accommodated industry, leisure, cooking, dining and sleeping and was often shared with livestock – sheep, goats, a pig perhaps. The little furniture they possessed might comprise a wooden table, some chairs and a wooden box as a cradle for the baby. In those without a chimney, the smoke from the stove found its way out through the roof and doorway. In huts lacking a stove, a fire was kindled on the floor for cooking. Outside was a simple water well, and toilet facilities consisted of a hole in the ground. They were difficult to heat, and in the rain could be damp and cold. With a high risk of infectious diseases, peoples' life expectancy was often relatively short, and babies and young children would die of pneumonia and tuberculosis. The cabins degraded quickly and needed to be patched up regularly. Contemporary images show some finely crafted and well-maintained dwellings while others appear as little more than dilapidated, squalid heaps of crumbling earth. The organic nature of these structures means that they left little trace on the landscape once abandoned.

A 1910 account of a Dutch plaggenhut dwelling offers an insight into the life of some of the last inhabitants on the Veluwe. A roof of heather sods; walls formed of half-rotten boards and old masonry stones; numerous holes and cracks in the walls that had been creatively plugged with a range of materials including cardboard and rags; the few windows had greenish, cracked panes; and a door hung crookedly on its hinges and was secured with no more than a piece of string. Hendrikje van Dijk, the 87-year-old woman inhabitant, worked a small area of arable and pasture – land wrested from the surrounding sand – with her one remaining son, farming sheep, potatoes and rye. Her life was a constant struggle against the sand, which during severe storms would accumulate halfway up the windows of the plaggenhut and infiltrate everything inside, including coffee, sugar and other kitchen cupboard items. She fetched water with a bucket from an open well on the heath. Nonetheless, despite living in poverty with a modest flock of only eleven sheep, the contemporary author described the household as a rich one, dwelling more contentedly in their 'lonely hut' with an open view and far from all worldly noise than the 'slave existence' of those living in overcrowded urban tenements in a turbulent world of worries.

In 1901 the Dutch government legislated on housing, introducing the concept of building standards. Building houses without a permit was prohibited, and municipalities were able to declare plaggenhuts uninhabitable and proceed with their demolition. However, the rate of new house building was slow, many hut residents did not wish to leave their makeshift homes, and they continued to be constructed until the 1930s in a few rural regions. Not until the middle of the twentieth century were the last plaggenhuts demolished; they were often burned because they harboured vermin – lice, fleas and cockroaches among others. The plaggenhut is a grim but also fascinating historic feature of habitation on the European deserts. These relict pre-modern dwellings were associated with pre-industrial land uses, modes of existence and societies, and persisted long after the advent of urban planning. The plaggenhut was a tangible element of a long tradition of independent and resourceful people in the margins of history with a strong desire to live on their own terms who successfully housed themselves in self-built dwellings. While there may have been many thousands similar across Europe by the end of the nineteenth century, mentions are unaccountably scarce in the literature. A handful of reconstructed plaggenhuts are exhibited in Dutch museums, and a few contemporary examples are available for rent as holiday stays. Kootwijkerzand is mostly unpopulated today.

DRIFT SANDS DECLINE AND 'REACTIVATION'

Economic changes commencing in the mid-nineteenth century had a major impact upon the Dutch drift sand tracts. The changes were driven by the introduction of industrially derived fertilisers and the collapse of the wool industry. Also, communal land use was abolished in an effort to improve agricultural prosperity and development, meaning that much of the large-scale communal landscape was enclosed in numerous small private plots each bordered by hedges delimiting property and controlling livestock. The enclosures resulted in some reduction in the extent of drift sand expanses, though most of the feasible reclamation had already been achieved. At the same time, as demand for wood for pit props for the mining industry increased, the Netherlands government began investing in large-scale afforestation of the drift sands from the start of the twentieth century. National policy supplanted local-scale action, and large tracts of desert disappeared under a canopy of trees. Tree planting proved an effective means of fixing ('stabilising') the unruly sands and minimising sand encroachment during extreme wind erosion events, but taken together the changes undermined the already declining economic rationale for the Dutch deserts' very existence. Scots Pine proved to be the tree most capable of surviving in the poor soils, but the dense new woodlands of this exotic species were of relatively low biodiversity value. By 2008, pine forests covered over 40% of the Veluwe. Planting also introduced to the deserts a prolific source of tree seeds of an alien species, the legacy of which would significantly disrupt the conservation of the habitat of these places in the future.

Plaggen harvesting activity declined on Kootwijkerzand with the introduction of industrially produced fertilisers. Grazing declined too, and open expanses of drift sand dwindled correspondingly. While at the turn of the twentieth century Netherlands drift sand areas covered up to 95,000 ha – approaching 3% of the country's land area – by the 1960s only 6,000 ha remained, approximately 4,000 ha in 1980, and in 2009 only 1,500–1,600 ha remained active. The redundant Dutch deserts might not have survived into the twenty-first century at all had Dutch national afforestation policy not been amended under pressure from conservationists seeking to moderate its impact, and in 1925 about 700 ha of Kootwijkerzand drift sand and heath was removed from the programme and designated as a nature reserve in recognition of its ecological and social significance. No management measures to maintain the area's primary interest were instigated at the time, however, which was unfortunate because, unlike most landscapes, Kootwijkerzand and other drift sand tracts are dynamic in nature, changing constantly in response to wind erosion and other natural processes. Most drift sand vegetation and fauna require bare sand and a certain level of erosive activity to survive. But sand drifts disappear without active and targeted management, for once ecological succession in the form of vegetation coverage on drift sand reaches about 40%, sediment transport halts almost completely. The inevitable result was a rapid diminution of the area of drift sand as pioneer vegetation colonised the expanses of open sand no longer being denuded of vegetation by plaggen or sheep. This invasive vegetation included seedlings from surrounding pine plantations and also a dense surface layer of algae, which can develop rapidly to form a crust a few millimetres thick which inhibits or even stops wind erosion, retarding drift sand development and accelerating natural succession. Not until 1960 was the first management strategy to conserve the drift sands produced. Now the remaining drift sand areas are protected as nature reserves. Nonetheless, the area of bare sand on Kootwijkerzand continued to dwindle, with 60% lost in the decade 2008 to 2018 as pioneer vegetation and self-seeded trees continued to spread over the site.

In a signal reversal of the tide of environmental policy, the Dutch have started actively restoring their remaining deserts! After centuries of efforts to suppress them, some of the Netherlands' former sandy wildernesses are being recreated. From around 1980 habitat management plans have aimed to restore former areas of open drift sands. Here, a passive approach to ecological management is not an option: in a policy highly significant for the Dutch deserts, *reactivation* of former open sand areas is being pursued, with the goal of expanding the area of drift sand. It is perhaps ironic that the best areas for optimal development of typical drift sand environments are the deflation plains, or blow-outs, those locations of extreme environmental conditions which were formerly managed by human

Dutch desert dunes with stumps of former trees.

intervention. The practical measures required to achieve the wind erosion activity upon which the drift sand ecosystem relies for its continuation are necessarily intensive and dramatic, involving the comprehensive clearance of vegetation in order to reactivate the wind erosion activity upon which the drift sand ecosystem relies for its continuation. In the twenty-first century, then, the Dutch heaths are once again being deforested on a large scale through the felling of conifer plantation trees. Grass, shrubs and the moss and litter layers must also all be removed, along with the uppermost layers of soil, stripping organic content and creating new active drift sand exposures in a process curiously echoing the plaggen method of past centuries. Maintaining a continuum of the various successional stages from bare sand to fully vegetated areas with gradual transitions between each is important to maintain biodiversity and minimise the risk of species loss. Ecological restoration activities are planned and managed at a fine level of detail – sometimes down to the scale of single square metres – and over timescales of 10–15 years.

What does this landmark reversal of policy look like on the ground? The scale and intensity of the required conservation management operations must be seen to be believed. Of necessity, ecological restoration needs to be undertaken on a large scale, because wind erosion cannot get going properly on small drift sand areas. Bare sand expanses need to be large enough to enable a sufficiently strong wind fetch thus ensuring effective erosion leading to sand transport. Wind velocity is inhibited in the zone immediately downwind from the surrounding forest, meaning that open stretches of several hundred metres in length are necessary to facilitate sufficiently intense aeolian activity. A threshold figure of 100 ha has been quoted as a working minimum. On-site, the casual visitor is confronted by a scene of devastation, with a dayglo orange-clad workforce employing monster machinery to rip up the landscape wholesale. Mature pine trees are torn down by fearsome machines within a matter of seconds, topsoil is removed using the sort of industrial equipment more usually associated with road construction projects than vulnerable nature reserves, and formerly wooded expanses give way to open, scruffy vistas of dusty yellow-white sand and very often relict dunes. For one reactivation project in the National Park De Hoge Veluwe, 15,000 pine trees were felled over an area of 130 ha. The litter layer was removed, and 10–25 cm of topsoil stripped from about 40% of the area to expose driftable sand in an intensive operation that yielded 18,000 m^3 of soil biomass residues. Sediment transport processes have been reactivated, though mostly occur only a few days each year during extreme erosion events.

Several drift sand tracts have been completely restored in recent years. Such large-scale reactivation projects have come at a considerable cost in terms of both financial and perhaps political capital – the former obtained from the EU and local government sources, because small-scale not-for-profit nature conservation NGOs do not have sufficient financial resources to fund such operations. The cost of comprehensive clearance was estimated at €20,000 per hectare at 2008 prices, though if tree stumps and topsoil were left in place this reduced to €5,000. A rotational habitat management strategy based upon the regular reactivation of areas has been employed, in which expanses of low ecological quality and with the potential of being reverted to bare sand covering about 20% of each drift sand area are identified, and clearance undertaken over a three- to five-year cycle.

Scots Pine on Tongerense Heide on the Dutch Veluwe. (Agnes Monkelbaan, CC BY-SA 4.0)

Conservation site managers must plan for a long-term commitment of resources, in support of a process which in nature conservation terms is arguably simply making up for a century or more of neglect. The Dutch deserts have gone from being revenue centres to cost centres; economic returns from conventional forestry operations are being forgone for an alternative set of returns brought by biodiversity conservation.

Physical processes once deemed a threat to human wellbeing are now being promoted as a matter of national policy. As pine plantations grown specifically to stabilise the invasive sands are felled, the desert is coming back! However, the modern-day task of turning back the ecological clock in this way is an expensive and long-term business. Ecological succession proceeds in the normal manner – for nature abhors a vacuum, and bare earth represents an opportunity to expand – meaning that habitat management work needs to be undertaken repeatedly. Nature is capable of outstripping the best efforts of humans, and simply to maintain the present extent of bare sand requires clearing an estimated 55 ha of forest annually across the Veluwe, a target which however remained unfulfilled over the decade 2007–2018. Flocks of mouflon sheep are kept to maintain the habitat, a sickle-horned relative of the primitive domestic sheep well adapted to dry heath and sand environments which is valued for its ability to browse the young Scots Pine seedlings which invade quickly and densely in these landscapes. However, the wolf has been recolonising the region, and this predator at the top of the food chain has been merrily reducing the numbers of sheep, deer and wild boar. Many ecologists welcomed this development, maintaining that wolves could help the ecological balance on the grounds that the former absence of natural predators had led to wildlife overpopulation necessitating the shooting of 50% of the deer and 80% of the boar populations to maintain acceptable numbers.

High-intensity management for biodiversity conservation is often controversial. The removal of trees on any scale is typically met with negative public reaction as the appearance of treasured local landscapes is quickly and radically transformed. Nature conservation approaches on the Veluwe entailing high-intensity methods have been criticised by Dutch advocates of wilderness and forest management based on natural processes. Real nature is not to be influenced by human intervention but should instead be dominated by natural processes; management focused on the preservation of certain species is not based on real nature but on the artificial preservation of a mainly cultural landscape through mowing, turf cutting and nutrient stripping, which has been parodied as mere gardening. However, Simon Schama maintains that the human world and the natural world have always been inextricably intertwined. Others agree that human influence is not all negative, citing a likely loss of biodiversity due to the disappearance of specialised wildlife species associated with cultural landscapes. A further issue is that heathland restoration projects promoting bare desertic soil and a depauperate flora and fauna sit a little uncomfortably alongside conventional habitat management strategies that typically aim to achieve maximum biodiversity and continuing natural vegetation succession rather than the opposite. Finally, in a district once almost under siege from invasive drift sand, there remains a concern that encouraging the capricious desert will bring a return to the bad old days of sand floods when wind erosion processes were essentially out of control. As with all biodiversity planning, it needs to be acknowledged that each site is unique, and critical judgements need to be made about the relative worth of the range of habitats achievable through intervention at each individual location, prioritising habitats with reference to their scarcity.

Other strategies to conserve the Dutch deserts involve more than direct habitat management measures. Leisure bungalow parks located in sensitive natural areas have been removed and fences dismantled; a former industrial estate built across a valley was relocated to unblock an ecological corridor important for wildlife migration; a disused military complex near Nunspeet was returned to nature rather than being redeveloped as a business park; and there are plans to close and remove some asphalt roads to minimise disturbance in the most important and vulnerable areas. The Dutch take biodiversity conservation seriously.

ATMOSPHERIC NITROGEN DEPOSITION

Not all factors behind drift sand succession can be addressed at site level, however. Air pollution has been identified as a key driver behind the decline in the area of actively moving sands over recent decades, and the ecological transition from bare sand ultimately to closed-canopy woodland is accelerated by. Since the 1960s at least a trend of increasing deposition of atmospheric nitrogen has expedited vegetation succession towards a closed forest as algae growth is promoted. A surface crust of only a few millimetres effectively stops wind erosion.[2] Nitrogen deposition also accelerates the growth of pine seedlings on drift sand areas, a further effect being a decline in the diversity of lichen species which may ultimately disappear completely. Invisible atmospheric nitrogen deposition accelerates the contraction of drift sand expanses together with populations of typical habitants like the Grayling butterfly. Dependent upon dry and sunny sites with sparse vegetation and patches of bare ground – and more usually found along the coast – the Grayling is another of the relatively few fauna species that can thrive in the extreme conditions of this landscape. Heath Star-moss, a dense yellow-green invasive alien species introduced by human activity, encroaches freely over these areas, its growth boosted significantly by atmospheric nitrogen deposition. Adapted to acidic nutrient-poor sand, once established on drift sands this bare-soils pioneer species can spread rapidly, swamping open areas, boosting biomass production and soil formation while also increasing landscape roughness, resulting in a reduction of bare areas open to wind erosion. However, stimulating aeolian activity through tree clearance and topsoil removal can retard its establishment, because moss growth is inhibited by the constant sedimentation associated with windblown sand.

So what is the source of the nitrogen deposition causing such problems for the survival of these Dutch deserts? Nearly half of all Dutch nitrogen emissions originate from agriculture, derived largely from animal manure as well as regular arable fertiliser. The Netherlands is a major centre of intensive livestock production, primarily pigs, with a pig-breeding herd numbering around 12 million, the highest livestock density in the world and the second-highest emissions of ammonia per hectare in the EU. Nitrogen evaporates into the atmosphere in the form of ammonia, infiltrating soil and groundwater through rainfall or via direct absorption from plants with consequent deleterious effects on the ecology of an environment where plant species are naturally adapted to a lack of soil nitrogen. However, pollution from excess nitrogen deposition is increasingly deemed unacceptable in a country seeking to conserve its natural environment and biodiversity and to comply with the EU Habitats Directive and the G7 Western governments' global targets to reverse biodiversity loss. Accordingly, it has been determined that improving the quality of drift sand habitat over the long term can only be achieved with a substantial reduction in atmospheric nitrogen deposition. Farmers are obliged to reduce nitrogen emissions, and among other measures the government plans to address the issue of livestock numbers, particularly in areas near important nature conservation sites. Dutch agricultural output may be significantly constrained to achieve conservation goals on their deserts.

In the late afternoon, I stumbled as I retraced my route across the unconsolidated sands of Kootwijkerzand. Picking my way through the trees planted to constrain this Dutch Sahara, I headed for the bus stop and returned to my airless budget hotel room

---

2   Koster, E.A. (2009) The 'European Aeolian Sand Belt': Geoconservation of Drift Sand Landscapes. *Geoheritage* 1(2): 93–110. https://doi.org/10.1007/s12371-009-0007-8

for the night. Though only a scant few kilometres to the west, Kootwijkerzand seemed far distant from the safe, solid and civilised town of Apeldoorn. The bus had transported me from a quintessential European desert – a chaotic and wild open space that once threatened to invade 'our world' – to managed and orderly territory, a realm of permanence and law and order.

Kootwijkerzand is a prize Dutch desert wilderness, a stunning, sprawling landscape feature stretching to the horizon – though one not easy to picture by those who haven't visited – and the Netherlands is home to some of the most outstanding and extensive examples of the European desert phenomenon. Once abandoned and fated to disappear under a canopy of conifer trees, the Dutch desert tracts are among the most important semi-natural cultural landscapes in north-western Europe, and their distinctive flora and fauna are now rightly recognised and appreciated as the national ecological family silver that they represent. Classic desert geomorphological features are here for all to enjoy, making their deserts yet more remarkable in the otherwise unspectacular Dutch national landscape. Only the most nutrient-poor Netherlands heaths remained intact during the twentieth century; only those with the least productivity historically have survived to be preserved as nature reserves. But they are being conserved for more than simply their ecological and topographic features: they also constitute a precious fragment of the Netherlands' socio-economic and environmental history and culture, one which has been

Heath Star-moss, also known as 'tank moss', is an introduced pioneer species which will grow in abundance on open tracts after soil removal. (Michel Langeveld, CC BY-SA 4.0)

thoughtlessly neglected in many other countries. Theirs is a paleolandscape, for on the remaining tracts one can take in views that early humans would have recognised, unlike almost anywhere else in Europe. They are relics of pre-industrial times, bypassed by progress and manifesting a semi-wild edge dating from the times before most of the rest of Europe's land had been conquered, civilised, regularised, managed and made to look and function much like everywhere else. Being so quintessentially different from their regional surroundings, the Dutch deserts are heavy with mystery and their own unique sense of place, evoking a sense of awe. Here is a very different formulation of countryside and indeed the planet, loaded with a romance of imagined spirituality and set under big skies with oft-dramatic weather. There is an energising feeling of freedom here, a distinct air of liminality, and the weirdness of sand dunes located far from the sea.

The big news from the Netherlands is that the country has been reactivating its deserts, with the goal of expanding the area of drift sand habitat. The old enemy – drifting sand – is being unleashed once again. In recreating their formerly suppressed and disparaged sandy wildernesses, the Dutch have displayed the political will (and committed some of the required resources) to conserve some of the most outstanding and threatened instances of the European desert wilderness phenomenon. In such an intensively developed country, the Dutch have little nature left to conserve, and with their remaining deserts they have something relatively unique. With its approach to ecological management, the Netherlands is one of the leading countries in the conservation of Europe's deserts. An amusing irony regarding their conservation efforts is that, in many ways, they are simply replicating the land-use changes of Neolithic times: felling forests, destroying much or all of the vegetation and creating wastes – but this time with power tools. If God created the Earth, then the Dutch created the Netherlands – but not just the once water-dominated bits but the arid bits as well. This is a country of unsuspected environmental extremes: from the aquatic to the desertic. Perhaps it is ironic that, to retain something like wilderness in an intensively developed country every last square metre might need to be planned and micro-managed in perpetuity. But while conventional habitat management is critical for the long-term survival of the incomparable Dutch sandy wildernesses, the less tangible and much more controversial atmospheric nitrogen deposition issue may yet prove to be an intractable one.

Chapter Four

FRANCE: THE DESERT THAT DISAPPEARED

Balmy from spring equinox to autumn equinox, southern France is a wonderful place to be. In the south-west corner, the land around the handsome regional capital of Bordeaux produces some of the finest *vin* in the world, of which St Emilion, Pomerol and Sauternes are but three favourites. The largest and most important wine-producing region in France, the Bordelais has about 125,000 ha under vines, and 2019 estimates indicate that a single hectare in the Pomerol appellation was priced at €2.3 million. Yet only 50 km to the south-west lies a vast, infertile European desert covering ten times that hallowed area, flat as a crêpe, almost entirely uncultivated and with a physical environment thoroughly dominated by sand. A vast, empty and mostly unknown region whose story has only rarely been told, yet once one of the largest deserts in Europe. For hundreds of years, its thinly scattered population practised an ingenious agricultural system to eke out an existence on the desperately infertile soils of Les Landes – the desert of Gascony. The inclusion here of Les Landes knowingly bends some of this book's working criteria for distinguishing European deserts, but the 'French Sahara' represents such a prize instance of the phenomenon that it had to be included.

In 1800, the south-western corner of France was said to be the most barren district of the country. Far from Paris, the people of the Les Landes region between Bordeaux and the Pyrenees had a reputation for their peculiar way of life. They were regarded as living in a world fundamentally different to that of most other French people – meaning worse. While elsewhere much of *la belle France* epitomised the green, carefully husbanded and productive environment of rural western Europe, the Landes region was referred to disparagingly by numerous authors as desert. The Landais people were regarded as poverty-stricken unfortunates living in a hellish, remote and disease-ridden region who had by necessity to cope with a sobering litany of existential risks largely unknown elsewhere. Today the desert has all but disappeared, and Les Landes is integrated into modern French society with its economy transformed and its people no longer poor. But while most of the former existential threats including dread disease have been consigned to history, few or none of the basic environmental fundamentals have changed here since the nineteenth century. While mostly obscured from sight today, the desert persists, and it's *huge*.

LES LANDES

Les Landes is a vast, flat plain, a sandy desert bordering the Atlantic coast by the raging Bay of Biscay. There are few comparable environments on such a scale in western Europe. Think south-west France, practically Spain. Imagine the longest, straightest and least-developed coast in western Europe. Immediately behind the endless sandy beaches a line of dunes run parallel to the coast, a full 230 km south from the mouth of the Gironde estuary – Bordeaux's river – almost to the Spanish border. Now picture a vast territory stretching inland from this coast, forming a recumbent V-shape tapering to its apex near the small town of Nerac a full 130 km to the east. This, the near-equilateral 'triangle of the Landes', is a major region immediately north of the Pyrenees, yet one about which most people know nothing. Even to the French, Les Landes is an unknown region. A brief glance at the map reveals... very little, over an area spilling across three *départements* and covering a full 2.5% of mainland France. The few major roads and railways are dead-straight because there were no natural physical obstacles to swerve around or tunnel through, and there are almost none of the handsome castles and chateaux for which France is rightly famous. Few people live here, which is curious given its high annual insolation and summer temperatures frequently rivalling those of the Mediterranean. Summer daytime temperatures can vary wildly, with a temperature range of 30 °C or more between sunrise and mid-afternoon, and the region regularly experiences summer drought.

A small and sleek raptor, the Black-winged Kite is known for its habit of hovering over plains and open grassland seeking its rodent prey. (Stephen Temple, CC BY-SA 2.0)

The herbivorous European Brown Hare lives in a variety of open habitats including desert, steppe and unimproved grassland, feeding mainly on grasses and herbs. With long, powerful limbs it relies on high-speed endurance running to escape its natural predators. (Caroline Legg, CC BY 2.0)

This former open desert comprises a huge plateau covered with a sand sheet. The term plateau is more usually applied to rocky upland areas, but the Landes plateau is lowland and flat, with only the barest slope (≤ 0.5%) tilting this major landform westwards towards the Atlantic coast. Uniform across the region, the quartz sand of Les Landes is distinctive for being coarse-textured, strongly acidic and so intractably infertile that it has merited its own name: *le Sable des Landes*, the Sand of the Landes. Partially because of its relative youth in geological years, Les Landes is poorly drained, with no major rivers over large areas of the plain. About one-third is dry: very dry indeed, the friable and free-draining yellow soil having little water retention ability. The remaining two-thirds is underlain at a shallow depth by small patches of dense and largely impermeable layers of hard pan, locally termed *alios*. These layers of sandstone, cemented by organic compounds and iron, are hard to penetrate, severely limiting the ability of plants, crops and trees to put down deep roots. Also, the layers above the *alios* have a very low water reserve leading to a water deficit in summer as the soil dries out, while winter rains cause the formation of a temporarily elevated water table clogging the soil above the *alios*. Some areas regularly become swampy despite the loose sandy nature of the soil, leading to the curious phenomenon of seasonally waterlogged tracts of wet desert – *désert humide*. Here is where we knowingly bend some of the book's criteria for identifying Europe's deserts,

Bracken is one of the most familiar heathland plants. Also known as eagle fern, it grows from an underground rhizome forming dense colonies of genetically identical fronds. (Hans Hillewaert)

Route d'Escourssole à Labouheyre image by Félix Arnaudin – 1880 – typical historic Landes panorama. (Félix Arnaudin, Public Domain via Wikimedia Commons)

for not all of Les Landes is year-round dry desert. But this is nothing new: scattered iron pan layers are frequently present beneath the sandy, acidic podzol soils upon which many European deserts are located, causing seasonally waterlogged areas of *désert humide* despite the free-draining nature of *le Sable des Landes*. To repeat: the French Sahara so epitomises the European desert phenomenon, it must be included.

In the agrarian age, the appearance of the Landais Desert was uniform across the region. There was nothing dramatic to be seen: the landscape was mostly flat, predominantly open, with few or no fences and monotonous to the horizon in every direction. Such unvarying terrain on such a large scale is uncharacteristic in small-scale Europe. The former aboriginal forest had been heavily depleted, most notably between the fifteenth and eighteenth centuries to feed the demand for wood for construction and fuel, and to provide land for sheep grazing. Here and there were small plantations of pine, or *pignada*, battered by the incessant breeze, and yet less frequent farms, often with a few oak trees to provide shade and acorns for pigs. Les Landes translates as 'The Moors', and few floral species could survive here. The dominant prospect was of vast expanses of extremely rough land supporting depauperate heath vegetation among areas of bare ground, exposed sand and gravel often devoid of higher plants. Sparsely distributed grasses mingled with mosses and lichens, broom and gorse bushes, and ragged low-growing dwarf heather evergreen shrubs. More moist areas supported Purple Moor-grass, a tough,

Sheep's Sorrel is a flowering plant of the buckwheat family with maroon flowers. Commonly found on acidic heathland, it is frequently one of the first species to recolonise disturbed areas such as the sites of fires and abandoned mining sites.

fibrous perennial, and Bracken. Low-relief sand dunes are an occasional feature of this landscape, aeolian geomorphic forms produced by accumulations of sand and dust found in deserts the world over. Typically up to two metres in height but ranging up to ten or more, with some extending for several kilometres, these are oriented according to the predominantly westerly wind direction. Dune profile shapes include the classic barchan (crescentic), the parabolic, the transverse and the dome. Mostly lying beneath a canopy of trees, these classic desert geomorphological features may today be concealed – but they persist nonetheless.

HISTORICAL LANDES FARMING

Historically the Landes region was renowned for presenting an exceptionally difficult prospect for farming. How did the Landais people manage to subsist and perhaps prosper in such a low-productivity environment? In the agrarian age, most people were *paysans*: peasants living in the countryside, their livelihoods wholly dependent upon subsistence agriculture. Each year, they needed to grow sufficient food to feed themselves until next year's harvest. Failure risked food shortages, malnutrition, disease, conflict, famine, starvation and death. Good-quality soil was the critical resource supporting this ceaseless task, but there was and is none in this desert; so instead it had to be fabricated. Let's take a few moments to understand the ingenious system of mixed farming historically employed and refined by the Landais *paysans* to survive in these desert wastes.

Plants are the basis of all life, and bread was the dietary staple of the French rural poor. Today we perhaps fail to appreciate just how central the humble loaf was to diets in the agrarian age. Every village had a bakery. Bread was made from grain grown locally over an annual cycle, ground into flour at the local mill and baked daily in a communal oven. This demanded a reliable supply of flour from wheat or rye grain. Of all the cereals, wheat is perhaps the most valued for its nutritional qualities, digestibility and arguably flavour; but it requires fertile soil to flourish. A relative of wheat, rye was widely grown for bread in France, being less demanding, more resistant to disease and often growing best on sandy soils of low fertility. Its fibrous straw was valued for use as livestock bedding, too. However, conventional arable agriculture as practised over much of the rest of France was not an option in Les Landes due to the poverty of the soils: the desert soil of Les Landes was unable to support even rye cultivation. How could the Landais people grow rye in a European desert?

The key to the soil fertility issue was nitrogen. Nitrogen is primarily responsible for supporting plant growth from the earliest stages of shoot growth through to stems and new leaves, though it is also the nutrient most usually deficient in agricultural soils. Twenty-first-century farmers routinely dress their fields with nitrogen in granular form, but in the agrarian age such fertiliser was unavailable: livestock manure and other organic wastes were the sole sources of nitrogen for sustaining crop yields. The task of the Landais people was to ensure a reliable supply of nitrate-rich livestock manure – *fumier* in French. Successful production of rye required enormous quantities of manure for application to the rye-growing fields. Nitrogen losses from the soil necessitated a continuous input of *fumier* to maintain fertility, without which yields declined rapidly. Livestock manure was produced, collected and spread upon intensively farmed cereal plots producing rye.

The Landais farm, or *airial*, was centred upon plots of land devoted to cereal growing. The tasks required to prepare the seedbed to maximise harvest yields had of necessity been raised to the level of a fine art. Rye production was the main focus, confined to relatively small, compact plots each of perhaps three to five hectares around the villages. The few contemporary accounts make clear just how much effort was required to maintain the productivity of these intensively farmed areas, whose appearance contrasted dramatically with the surrounding sandy wastes. In autumn, each plot was fashioned into rows of deep ridges with the aid of a livestock-drawn plough, after which all subsequent operations were accomplished by manual labour. Rye seed was sown on the tops of the ridges to preclude any potential issues with winter rains or, in some locations, floods and waterlogged soil. The land might also need to be top-dressed with straw to mitigate the effects of heavy winter rain. In spring, another crop was sown in the furrows between the rye ridges, typically millet, or maize which could survive the rigours of the heat in summer and produce a second harvest from the land. Attentive weeding minimised competition for precious soil nutrients, and keeping birds and other pests at bay was another necessity. Come summer, when the seeds were ripe the rye was harvested: in a labour-intensive process, the plants were cut down, then threshed to separate the grains from the straw, and finally winnowed, when their outer husks (the chaff) were separated and discarded, before storage, milling and subsequent food use. Post-harvest, manure

The Écomusée de Marquèze is an outdoor museum recreating a typical nineteenth-century Landais farm community with livestock and translocated buildings.

diligently collected from the sheep was spread liberally over the arable fields – by hand, of course – to fertilise them before the annual cycle recommenced. Unlike common practice elsewhere, where fields might be allowed to lie fallow for a year or more after harvest to preclude soil exhaustion, the practice of 'resting' the land in this way was unknown in Les Landes because the intense management regime applied to the plots enabled continual cultivation.

So, the key to human survival in the desert environment of Les Landes was… *fumier*. Sheep manure was the source of the nutrient stream vital to support the rural population. Composted sheep manure provides fertiliser and acts as a soil treatment, but it is bulky and awkward to acquire, transport and spread. The industrially manufactured nitrogen fertilisers of today are easy and clean to handle in white granular form, and at 25–30%+ their nitrogen value dwarfs that of sheep manure with a typical concentration of less than 1%. But in the agrarian age *fumier* was the only option for the Landais, and they needed an awful lot of it. The ultimate source of this vital nitrogen was the desert. The sparse and slow-growing plants able to survive here locked nitrogen in their tissues. The primary role of the sheep lay in nutrient scavenging and concentrating. Sheep grazing upon the plants ingested and processed this vital mineral, converting the sparse vegetation of the Landes desert into something usable: agricultural fertiliser. This was an agro-pastoral system dependent upon the production of livestock manure to support cereal growing: the sheep were the foot soldiers, and naturally shepherding was a vital occupation. In 1850 there were up to one million sheep making the best of the slim pickings available over the sandy wastes, and they were kept primarily for their faeces. Elsewhere in France sheep were raised for their meat and wool as much as their manure, but in this European desert their primary role was as fertiliser generators, and the communal wastes of the Landes were the inhabitants' most precious resource.

The concept of nutrient transfer underlay the economic system of the Landes. It functioned through a transfer of chemical fertiliser from the extensive desert areas to the intensively cultivated plots. The flocks roamed widely over the communal desert, moving up to 20 km each day as the paltry available herbage was progressively depleted in each successive location. The sheep were herded together and penned into a sheepfold, typically overnight, where the manure they produced could be concentrated in one place. To assist its handling, heather, gorse and bracken were cut and heaped in the sheepfold to be trampled and mixed with sheep excreta. *Fumier* gleaned in this laborious fashion was subsequently composted, excavated and transported to the *airial* by horse and cart before being applied to the rye-growing plots.

The mathematics of the nutrient-transfer formula is staggering. A commonly quoted sequence of figures provides an insight into the historical Landes human food chain:

- 1 kg of rye bread fed an adult, and a family of eight to ten people would consume 4,000 kg of rye bread each year
- 3,200 kg flour is required to make 4,000 kg bread
- to produce 3,200 kg of flour, 4,000 kg of rye grains must be milled
- producing 4,000 kg of rye grains required four hectares of land in the rye plots
- fertilising four hectares of land required 60 tons of manure

The Strawberry Tree is an evergreen distributed in France across the south and west regions. The fruit, a red berry with a rough outer skin, reaches maturity over a period of 12 months, coinciding with the autumn season and the subsequent flowering phase. (Manuel Torres Garcia)

- producing 60 tons of manure required 100 sheep; and
- feeding 100 sheep required 100 hectares of Les Landes, approximately the equivalent of 150 soccer pitches.

Truly this was a low-productivity environment. These figures make it easier to appreciate the scale of the challenge involved to survive in the French desert, and the precarity of existence there. The few people who lived there were entirely dependent upon their flocks. There were few or no prospects for increasing productivity in such an environment; a mixed farming system centring upon sheep was the only option. With precious little foliage for the sheep to graze upon on the barren Landes, a single sheep per hectare was all that this impoverished environment could sustain, a desperately low stocking density. Sixty tons of sheep manure needed to be generated, collected, transported and then shovelled onto the precious rye-growing plots. And all without mechanical means: the sheer scale of human effort required can only be guessed at, in a tedious and unrelenting cycle of *fumier* shovelling. This intensive cereal production would have been primarily for subsistence purposes, too, food for the immediate needs of these mostly self-sufficient peasant communities rather than as any form of cash crop.

Historically, farming in the Landes was a form of the infield-outfield cultivation system thought to have been widely practised on the European subcontinent's sandy lands and elsewhere. Based around the transfer of plant nutrients, the system operated at two different levels of intensity: an intensively farmed infield zone around the village,

Native to the Mediterranean and southern Europe, *Pinus pinea* is better known as the Stone, Umbrella or Parasol Pine. It has been cultivated for its edible pine nuts since prehistoric times. (Jebulon, CC0, via Wikimedia Commons)

the best land which was under permanent or near-permanent arable cultivation; and an extensive outfield zone of lands distant from the village which was exploited at a low intensity. The practice of infield-outfield farming is a pre-industrial one, offering a means of cultivating soils in locations where abundant areas of marginal quality land are readily available. Forms of the system have been employed throughout the world for centuries, including Europe from the early medieval period at least until the start of the twentieth century, though the infield-outfield system may have ancient origins.

Remarkably, neither the desert sheep's meat nor their wool were of significant value in comparison to their excrement. The main function of the sheep was to serve the demands of intensive rye and millet production on the poor Landes soil, hence the term agro-pastoral. The sheep were not raised primarily for butchery, apart from a few young lambs sold in nearby towns; otherwise, the sheep meat was consumed locally. Shorn in June or July before the heat of the summer, the wool was sold to merchants who sent it to make coarse fabrics and blankets. The sheep's milk was left for the lambs. The few other Landes farm products included honey and wax from beehives, resin produced from pine tree sap – of which more later – and timber. In addition, each airial would keep pine trees for a harvest of pine nuts, *pignons de pin*. Familiar all around the Mediterranean for its softly beautiful outline, the seeds of the Stone or Parasol Pine were extracted from the cones for use in cooking, cakes and confectionery.

SHEPHERDS ON STILTS

The Landais community food chain was heavily dependent upon the shepherds, both men and women. The role of the shepherds – the *bergers* – was to take charge of the grazing, caring and breeding of a flock of 120–150 sheep, usually for someone else, a rural bourgeoisie of landed property owners descended from the peasantry who owned almost all the herds. They maintained the land's meagre productivity through the practice of the *burle*, regularly burning tracts of older woody vegetation, which brought fresh and tender foliage for grazing. With one sheep or fewer per hectare scattered across the desert plains, keeping the flocks under continual surveillance was problematic, so to keep track of their sheep the shepherds used a novel means: stilts! More usually familiar as children's playthings, those historically used by the Landais shepherds were *des échasses*, pairs of specially made wooden stilts that raised the shepherd a metre or more above ground level, as skilfully represented in the oil painting *Bergers landais sur échasses* by Jean-Louis Gintrac (1808–1886). Pairs of stilts were fastened around the leg below the knee by a tight leather strap – leaving the hands free to undertake essential tasks – with the *berger*'s weight supported with a wooden footrest. A third piece of wood, an elongated pole, was carried, enabling an ingenious tripod-like arrangement of stilts and pole upon which the shepherd perched effortlessly at a commanding height above the desert. Looking like some sort of overgrown long-legged insect and with one or two dogs in tow, the shepherds perched aloft for hours supervising their woolly charges, easily spotting wayward animals and potential threats – and, as the wolf was never far away, they routinely carried a gun, too.

Les Landes shepherds on stilts. (Félix d'Aquitaine, Public Domain via Wikimedia Commons)

The wolf is a devil of folklore, and while this apex predator is naturally shy and elusive they have been known to attack humans, particularly where they have become habituated to them or their natural prey is scarce. (Konrads Bilderwerkstatt, CC BY 2.0)

The stilts were also a means of transportation, turbo-charging the Landes peoples' ability to travel across the desert. Long strides enabled them to cover great distances across the sands with a practised ease while avoiding thorny shrubs and marshy tracts. So efficient were the Landes stilts that a skilled user was said to be able to speed along at the rate of a trotting horse. Women and children were skilled stilt-walkers too, and the postmen used them to travel between the thinly scattered settlements in a region with almost no transportation network. The most famous of all was M. Sylvain Dornon, the 'stilt walker of Landes', who in 1891 determined to visit the Franco-Russian Exhibition in Moscow. On his 1.2 m high stilts. Dornon left Paris on 12 March and reached Moscow 58 days later having travelled 2,800 km entirely on stilts, regularly covering 60 km a day. Walking on stilts was still an everyday sight around the turn of the twentieth century, and today they have become a symbol of the region, with a revival of folk culture bringing stilt dances, games and races to village festivals.

LES LANDES HAZARDS

Les Landes was a region of risk for its inhabitants in the agrarian age and remains one today in some respects. Meat was only a limited element of the diet, and most people were wholly dependent upon their subsistence crops of rye, millet and maize to survive. Yet the desert could be cruel and if anything in the food production chain went wrong, the consequences could be serious indeed. Usually a moderate and humid temperate oceanic regime, the region's climate nonetheless constituted an uncontrollable environmental variable, threatening hazards including wild Atlantic storms, sudden winter cold snaps, and regular summer drought. In 1868, the Landes food supply slid into crisis due to intense drought, leading to 'great misery' across the region by 1870.[1] In 1871, the drought was so severe that the rye harvest failed completely, and even apicultural products dwindled when the bees had no means of sustenance as even the hardy heather plants failed to flower. Described as another mark of general misery, theft of crops became a rural scourge during the four worst years, and the cultivation of rye and corn in the fields remained at risk until 1872 when harvest volumes recovered. Crop disease and pests were a constant source of food insecurity, depleting harvest yields and spoiling food stocks. Inter-village conflict could get rough when parishes sought to defend grazing lands against livestock encroaching from neighbouring communities.

An unhealthy place, Les Landes was notorious for malaria, which made 'sickly the unfortunate classes who vegetated there and of whom most died of fevers'.[2] The regional death toll included the town mayor of Lit-et-Mixe and his son over a 48-hour period in the nineteenth century. Not until around 1900 was the vector confirmed as the mosquito. Another common disease occurred here when, during times of poor harvests, maize was substituted as a dietary staple because it had a higher yield than other grains. Such a restricted diet was deficient in vitamin B3 – niacin – producing the nutritional disorder pellagra and its classic accompanying 'four D' symptoms: dermatitis, diarrhoea, dementia and death. Neither did the sheep thrive in the often very wet winters, frequently succumbing to diseases related to the damp conditions. Drifting sand was another hazard, an unpredictable force of nature with the ability to reduce agricultural land to infertile waste and swamp settlements within a few years. This was a particular problem along the Atlantic coast, where transgressive dune fields would penetrate inland over time, compelling the inhabitants to retreat and rebuild ports and villages. Today, the tallest sand dune in Europe lies on the western Landes coast immediately south of Arcachon. The gigantic Dune du Pilat is a steep 106 m high unvegetated burnished-gold monster made up of flyaway sand. Loved by paragliders and sand surfers, Pilat is 3 km in length and 616 m in width, and one of the few dunes in the world with its own web page. And, in keeping with Landes tradition, Pilat is invasive too, moving inland up to 5 m each year, slowly consuming forest, roads and houses.

In the past, Les Landes was notoriously risky for travellers. Dating back to the ninth century, French sections of the Christian pilgrim route to the cathedral in Santiago de Compostela in Spain pass through the region. Traversing Les Landes en route was dreaded by pilgrims walking the Camino de Santiago, who reported that they were frequently up to their knees in sand. A twelfth-century pilgrim's guide to the route described a desolate country where everything was lacking, for no bread, no meat and no

---

1   Traimond, B. (1986) Le voyage dans les Landes de Gascogne ou la traversée du Sahara français. *Études rurales* 103–104: 221–234.
2   Aldhuy, J. (2007) La transformation des Landes de Gascogne, de la mise en valeur comme colonisation intérieure (XVIIIe–XIXe siècles)? *Revue Géographique des Pyrénées et du Sud-Ouest: Sud-Ouest Européen* 23(1): 17–28.

The Dune du Pilat on the Landes coast (the sea is to the right of the picture). (Tanya Dedyukhina, CC BY 3.0)

sources of drinking water were to be found along the way. The region was described as 'insalubrious',[3] and its inhabitants reputed by some to be hostile to outsiders: a middle-class English woman who spent three days travelling through in 1854 was nonplussed when a male companion who bid a man cutting wood for charcoal 'a civil *bon jour*' was met with 'a most unceremonious salutation'.[4] 'Advancing with a very suspicious look, and wielding in a menacing manner a tremendously thick club', the woodsman queried whether the traveller was journeying alone. This question was interpreted as a possible prelude to a violent attack, so the traveller replied that his party numbered several people and spurred his horse on to make good his escape.

The skilled and hardy people of the Les Landes European desert were long disdained by outsiders. The region was remote, unknown, even perhaps feared, and the Landais frequently regarded as a race apart. 'The inhabitants of the Landes are a species of wild savages, in appearance, in mood, and their collective spirit' opined Lamoignon de Courson,[5] an aristocratic regional governor, in 1715. Eighty years later in 1798, Grasset

3   Cobb, C. (1910) Landes and Dunes in Gascony. *Journal of the Elisha Mitchell Society* 26 (3): 81–95.
4   Alethea, E. (1855) *Pignadar: Or Three Days' Wanderings in the Landes*. London: Longman, Brown, Green and Longman.
5   Aldhuy, J. (2007) La transformation des Landes de Gascogne, de la mise en valeur comme colonisation intérieure (XVIIIe–XIXe siècles)? *Revue Géographique des Pyrénées et du Sud-Ouest: Sud-Ouest Européen* 23(1): 17–28.

de Saint-Sauveur concurred, declaring the Landes peasants to be 'uncivilized; the kind of life they led made them quite rustic and almost wild'.[6] Thus, they were characterised as the Other, and as distinctly non-French: in 1810 the military doctor Jean Thore declared that 'the Landais form a tribe in the department apart and distinct from all the others, as much by their physique, their character and their morals as by their habits and the manner in which they live'.[7] To the Vicomte de Métivier, a first-time visitor in 1839, the Landais were in some ways a nation apart, organised in almost nomadic tribes.[8] Seven years later, Amable Tastu, a French poet and writer, noted in her 1846 *Voyage en France* that here she felt she was in some distant land quite foreign to her own civilisation.[9] A wild region begot a wild population.

Landais society was derided when in 1826 Baron d'Haussez opined that many things remained to be done before the civilisation there attained the height it had reached in the rest of France, a presumably carefully weighed and well-informed statement from the appointed prefect of the Landes. Lamoignon de Courson wrote of an 'almost uninhabited country, [where] one finds only houses, or rather scattered cabins in the countryside, very far from each other [...] We come across a house there only by chance.'[10] Perhaps unable quite to believe that the population was indeed actually French, observers identified overseas comparators for the Landais: no, they must be some foreign mixture of Celtic peoples with a Semitic or Arab branch, or Greeks, Bedouins, Tartars, Hottentots, or Indians. Again and again, an analogy between the inhabitants of Les Landes and other wilderness regions around the globe was employed in the eighteenth and nineteenth centuries, with savage Landais men and women depicted as 'rather under-sized',[11] stilt-walkers as 'wild and uncouth', and the silhouette of the motionless shepherd perched upon stilts described as looking like a sick heron.

French eighteenth- and nineteenth-century elites frequently compared the region to the world's deserts, deploring that these vast burning wastelands of endless horizons lacked any use. Tastu declared that Les Landes was like the deserts of Africa or Arabia,[12] and the cultivated and inhabited areas were likened to oases in a desert. Flamichon in 1816 wrote that 'on entering these vast deserts [...] the eye [is] amazed at an unexpected uniformity of appearance'; 'the spectator does not see the smallest unevenness of ground in the vast expanse of country with which he is surrounded on all sides [...] everything seems to be drawn up with a rule and a line. So nothing is more boring, nothing is more tasteless to the eye, than the aspect of the Landes de Bordeaux.'[13] The naturalist Bory de Saint-Vincent asserted in 1826 that 'the surface of the Grande Landes is subject to a

---

6   Grasset de Saint-Sauveur, J. (1798) *Voyage à Bordeaux et dans Les Landes où Sont Décrits les Moeurs, Les Usages et Costumes du Pays*. Paris: Pigoreau.
7   Thore, J. (1810) *Promenade sur les côtes du Golfe de Gascogne*. Bordeaux: Brossier.
8   Métivier, Charles-Gabriel-François-Hyacinthe-Denis (1839) *De l'agriculture et du défrichement des Landes*. Bordeaux: Th. Lafargue.
9   Tastu, A. (1878) *Voyage en France*. Tours: Ad. Mame et Cie.
10 Aldhuy, J. (2007) *La transformation des Landes de Gascogne*.
11 Cobb, C. (1910) Landes and Dunes in Gascony. *Journal of the Elisha Mitchell Society* 26(3): 81–95.
12 Tastu, A. (1878) *Voyage en France*. Tours: Ad. Mame et Cie.
13 Flamichon, E. (1816) *Théorie de la Terre*. Pau: Tonnet.

mirage which does not yield to that of the deserts of Egypt or from Arabia'.[14] Making a grand tour of France to study the country's cathedrals, the American Elise Whitlock Rose described the Landes around the turn of the twentieth century as 'bad lands';[15] a torrid Saharan region of waste and awesome terror; and a desert whose monotony was broken only by the terrifying phenomenon of sandstorms.

PROFOUND CHANGE

By the nineteenth century, recognition was growing that the distinctive Landais mode of life focused upon sheep, *fumier* and rye growing was a historic throwback, and one coming to an end. The Sahara of France, this great sandy desert 'must soon cover these last savages destined no doubt to disappear. Civilisation, in fact, drives them before it, as did in the United States, American colonization' wrote the historian Henry Ribadieu in 1859.[16] A substantial minority of the country's land area was essentially uncultivated – perhaps one-quarter to one-third – therefore adding little to the gross domestic product. Unproductive lands were considered ugly, frightening and dangerous, and Les Landes appeared to most observers as representing this 'anti-nature'. Starting in 1857, change was wrought on a vast scale, transforming the region comprehensively and forever. The Landes waste was to be cleared and made properly productive. A law of Napoleon III required that parishes improve their districts, meaning that, in effect, the whole region was to be planted with trees. The Maritime Pine, a hardy Mediterranean conifer able to survive in poor sandy soils, was planted on a huge scale; and rapidly, too, with one estimate indicating that by 1878 less than one-tenth of the former area of waste remained to be cleared.

By 1900 the desert was being extinguished as most of the region was afforested. Increasingly the remaining rye plots could not be fertilised in the traditional manner, and forestry employment was attractive. But a prize much bigger than mere wood was the resin produced by the Maritime Pine. The tree's high-quality sap was a feedstock for industrial processes manufacturing turpentine and rosin, increasingly valuable products at the time. Growing pines were each tapped by an incision made manually into the tree trunk and the yellow-white pine resin collected before being transported to local distilleries. *Le gemmage* became the main economic activity, and the new profession of *gemmeur* or *resinier* was a labour-intensive occupation, each man working about 7,000 pines to produce about 1.5 litres of resin per tree annually in what was described as a gruelling daily schedule. Despite the advent of this new rural industry, however, the population in the region remained scant. A vast network of ditches was dug to drain the more seasonally waterlogged areas, and new roads and even a railway were built, enabling forest products to be exported from the region. Numerous alternative crops and animals were trialled to

---

14 Bory de Saint-Vincent, M. (1826) 'Landes'. In M. Bory de Saint-Vincent, *Dictionnaire Classique d'Histoire Naturelle*. Paris: Rey.
15 Whitlock Rose, E. (1906) *Cathedrals and Cloisters in the South of France: Vol. II*. New York: G.P. Putnam's Sons.
16 Ribadieu, H. (1859) *Un voyage au Bassin d'Arcachon*. Paris: Tardieu.

The dark, long-tailed Dartford Warbler is the quintessential bird of the European deserts. It may be glimpsed as a small flying shape darting between bushes, or perched singing on top of gorse. (Vcebollada, CC BY-SA 4.0)

make the desert bloom – including, famously, North African camels, which were fêted as an ideal draft animal for forestry operations – but they almost always ended in failure. The Landais were quietly amused by what they regarded as carpetbaggers from Paris with their ill-informed efforts seeking to make a fortune in this infertile region. The buffalo were not suited to the environment, and alarmed the Landais sufficiently into thinking that they represented bad luck, so they massacred them. The last of the poor camels died around 1865, the often-harsh winters too cold for them, eaten away by humidity and decimated by tuberculosis.

The 1857 law changed the character of an entire region in a similar way as in many other desert locations across western Europe. Les Landes went from being perhaps the largest desert in France – or perhaps even Europe – to the largest human-made woodland in western Europe. The region is now dominated by near-enough *one million hectares* of forest, readily visible from space, 90% of which is a Maritime Pine monoculture. With its serried ranks of even-aged single-species trees, this is not at all like a natural forest. The loss of the region's former limited, though significant, biodiversity interest can only be guessed at. The soil tillage necessary to prepare the ground for tree planting would have decimated its habitats, compounded by light availability declining as the forest canopy developed. However, in an ironic twist, some of the best-preserved remnants are to be found in the military zones of the region, attracting breeding species including the Black-winged Kite – more at home in Sub-Saharan Africa, India and south-east Asia – Black Woodpecker, Red-backed Shrike, Woodlark, Dartford Warbler, Nightjar, Short-toed Eagle, Marsh Harrier, Hen Harrier and Montagu's Harrier. While the military zones are designated nature reserves, there are few other such reserves over this huge region.

Shepherds gradually lost access to the former wastes, and arson became one of the most prosecuted crimes in the Landes in the second half of the nineteenth century, as farmers and shepherds set the new pine forests alight. A landowner in La Teste decried this destruction of the forests that had increased the value of land a hundredfold by a 'handful of primitive savages' who clung desperately to their communal pasturage, bemoaning 'Is there nothing one can do so that civilisation may advance through the sap of the trees?',[17] but the tide of history was only going one way. Many farms were abandoned or burned down. The pine came to dominate Les Landes and its people in what was interpreted by some as a process of internal colonisation. Some lamented the upheaval of the Landes ecosystem, regarding it as the intrusion of capitalism into a society which until then had remained poor but relatively egalitarian. The landscape changed completely: instead of settlements surrounded by empty desert and distant horizons, the Landais people increasingly lived in clearings surrounded by forest. Today, forest fires are a hazard for the Landais, for Maritime Pine plantations are flammable and prone to high-intensity fire, not least because the sap includes hydrocarbons that are akin to those in petrol. There are about 1,000 fires annually across the region, with the 'Great Fire' of 1949 perhaps the biggest, striking during a summer of severe drought and burning 60 houses – killing 83 people including 29 in Canéjan at the gates of Bordeaux. 50,000 ha of forest was destroyed, an area nearly the size of urban Manchester. Storms are a continuing menace too, with those in 1999 and 2009 snapping and sundering the forests on a large scale.

Inevitably, cheaper fossil-fuel-based substitutes including white spirit superseded pine resin in the mid-twentieth century. The last resin factory closed in 1962, and commercial resin production ceased in 1992 when there were only 76 *resiniers* left. Growing pines for wood products has remained profitable, however, and now the forests produce wood and timber on a grand scale, with yields boosted by fertiliser and mechanisation. Today this industrial plantation forest employs 34,000 people, producing wood for paper pulp, packaging, furniture, building, panels and frames. The former Landes European desert has disappeared almost without trace; there are no commercial sheep farms, and few first-hand accounts of the former world there. However, unlike so many other European deserts which were not considered worthy of scholarly attention, Les Landes was a subject of fascination for its own chronicler. Félix Arnaudin (1844–1921) was a local-born writer, folklorist and ethnologist from a modest background who patiently documented the work and lives of the Landais people in writing and thousands of insightful monochrome photographs. Showing skilled and hard-working people living in a harsh desert environment, Arnaudin acknowledged the uniqueness of this former European wilderness and recognised that it was disappearing before his very eyes. His efforts were derided by his peers, who considered him an eccentric and nicknamed him Lou Pèc – the madman, in Gascon. Yet Félix was vindicated: today his work is archived in regional museums. In recognition of Arnaudin's lifelong work, modern Landes writer

---

17 Temple, S. (2009) The natures of nation: negotiating modernity in the Landes de Gascogne. *French Historical Studies* 32(3): 419–446.

FRANCE: THE DESERT THAT DISAPPEARED

Les Landes forestry operations – clearfell near Azur.

Tract near Azur cleared for tree replanting – relict dunes in the distance.

Jacques Sargos said that he had paid tribute to a country that had been slandered, and which would have been forgotten without him. Would that there had been more like Arnaudin in deserts elsewhere in Europe.

No one alive today knows what the Les Landes of the agrarian age was really like. Contemporary observers noted that, before the trees, the empty horizons here engendered a feeling of infinity, though the region also evinced a strong sense of loneliness and melancholy. A grandeur and intensity of desolation invested Landes scenes with a sad, solemn poetry peculiar to this region. The atmosphere of this still-remote region is uniformly mute today: quiet but for the soft roar of the breeze in the trees. Between scattered, depopulated small settlements with long, long roads in between, Les Landes feels airless, smothered by grey-trunked trees. No vistas, no perspectives, little colour and the same everywhere you look.

The Landes desert has been annihilated, though it took massive and sustained state intervention to accomplish it. Its passing marked an incalculable loss of environmental, biological, landscape and cultural diversity. Two formerly dominant ways of life have ceased, and there is almost no trace of their history. Tourist guidebooks gently advise that there are no significant cultural attractions on the plain, although the coast is attractive for the more adventurous beach-lover. The old men play boules in the village squares. Les Landes remains at one end of the French 'empty diagonal' or *diagonale du vide*, a wide hyper-rural slash across the country from north-east to south-west characterised by low population densities and youth outmigration. Although still low, the regional population is slowly increasing as some parts of this undiscovered region become attractive for

Twenty-first-century Les Landes – dominated by grey-trunked trees, near Lacanau.

Sparse vegetation regrowth on cleared forestry site near Azur.

second-home purchasers. In an ironic reversal of fortunes that is also quietly happening in other European deserts, what was once an undesirable desert wilderness is gradually becoming gentrified.

But the desert is still there. The physical elements of geology and soils may today be mostly obscured by a monotonous mantle of pine trees, but while humans have adapted their culture and economy to the desert conditions to some degree – as they have in deserts around the world – the fundamental dominance of the natural environment remains a given in Les Landes. While mostly out of sight today, this European desert persists, and it's *huge*.

Chapter Five

PUSTYNIA BŁĘDOWSKA: THE POLISH SAHARA

Kraków is one of Europe's most compelling medieval towns. A stroll around this former national capital confirms that it is the only major city in Poland to have survived the destruction of World War II largely undamaged. Dating back to the seventh century, Kraków boasts numerous historic houses, palaces and churches, ancient synagogues, a large market square and a Gothic cathedral. In 1978 UNESCO designated the entire Old Town for inclusion in its first World Heritage List. After enjoying the warm atmosphere of this handsome central European city, I elected to make a comparison. A tram ride out to the east of town opened up some very different vistas, the trip presenting as sobering a juxtaposition over a short distance as could be imagined anywhere – because thirty years prior to the UNESCO designation, the Polish government had selected genteel Kraków as the location for a giant new steelworks. Nowa Huta – literally 'The New Steel Mill' – became the biggest steelworks in Poland, and a new town of 50,000 people was grafted onto the historic city to house the workers. However, this large-scale industrial district was subsequently to contribute to turning the whole subregion into one of the most polluted in Europe, threatening public health as well as architectural treasures. It wasn't too long before I clambered onto a tram and headed back to Kraków.

A 50 km drive to the north-west of civilised Kraków yielded yet more unusual vistas: the greatest expanse of unruly desert sand in central Europe. Popularly known as the Polish Sahara, Błędów is a vast unvegetated desert set in the east-centre of the subcontinent, a landscape of bare sand and arid grassland. How did a desert develop here, in the south of a country not regularly associated with aridity? Local legend has it that the desert was created from sand from the Baltic Sea. It seems that underground-dwelling devils and demons became infuriated by the intrusive and noisy operations of the miners from the nearby small town of Olkusz, who were stealing their silver. One fiendish creature decided that the humans' mines should be blocked up, and determined to do so by flying to the beaches of the Baltic coast – 500 km away in the north of the country – with a huge sack to collect sand for use as filling material. However, on the return trip the sand sack snagged on a tall local church tower and the bag ripped open, whereupon huge quantities of sand spilled over the fields thereby creating the desert at Błędów. Modern-day geologists seem not to be in agreement with this imaginative geological creation tale, however.

PUSTYNIA BŁĘDOWSKA: THE POLISH SAHARA

Sand and sand dunes cover much of Poland, occurring as single and small sets of dunes and extensive continuous fields. Everywhere else, these sand sheets are covered with vegetation – mainly pine forest, for more than one-third of Poland is under woodland – but Błędów is a scarce instance of exposed sand. Buried deep below, the solid geology of Błędów is limestone, which was covered by sand deposited by glacial meltwater and outwash floods during glacial times in the Quaternary period. This was no light dusting, for it is reckoned to comprise about 2.5 billion cubic metres of sand, the depth of which varies from an average of 40 m to 70 m or more. Błędów is a huge pile of sand by any standards. Over millennia, the wind would have got to work episodically on the extensive sand surface that was formed, as doubtless also did early humans, but equally episodically the proto-desert was invaded by trees. A wooded landscape prevailed until early medieval times, when in the thirteenth century mining operations commenced on a large scale around Olkusz.

On stepping out onto Błędów today, one's immediate impression is just how egregiously out of place the desert looks here. The Polish desert resembles a gigantic, almost infinite sea of sand. Set at an altitude of about 350 m, Błędów is an arid and barren environment hostile to both plants and animals. Hemmed in with trees, the

Pustynia Błędowska – the 'Polish Sahara'.

characteristic vegetation of southern Poland, the desert's margins are sharply delineated, mostly lacking any intervening scrub vegetation stages and making the whole area look weirdly unnatural, as though it was excised cleanly out of the woods with some monster knife. Here it is unnaturally bright, dazzling even, as the sun's rays reflect vividly from the desert surface. Unwelcoming, unsettling. No one lives here. Parched: not a drop of water. Wind dominates. Vistas are wide and sweeping. Skies are huge. Colours are grubby white, pale yellow, grey and dull brown. Tracks across the desert are ill-defined, wildly branching and arduous to traverse. The southern half presents a gently rolling rippled surface of low-profile dunes, with numerous stretches of bare sand separating thinly vegetated areas supporting pioneer grasses and a few plucky tree seedlings. By contrast, the northern section, naturally subdivided from the southern by the valley of the Biała Przemsza river, is almost completely level and horrifically barren – a huge expanse of bleached-white flyaway sand.

MINING HISTORY

A key characteristic distinguishing Błędów from most other European deserts is its valuable mineral resources. Few deserts around the world have these to offer: fertile soil, water, wood, rock or other building materials, or minerals. Sand and gravel have their uses, as attested to by the proliferation of historic quarries on many European deserts, but these minerals are high-bulk low-value products destined primarily for local markets. But Błędów is different, for it is underlain by a very significant economic resource: galena, a metallic mineral from which is derived lead, zinc – and silver. Galena is a naturally occurring ore associated with limestone rock, and has been mined here since the Late Neolithic, bringing prosperity to an otherwise unpropitious rural area. Small amounts of silver occur with galena deposits, a coveted material used in status display for time immemorial and destined for national and international markets, unlike the mere functional commodity that is lead. Silver's value far surpasses that of the more common metal – over 300 times more, in fact. The galena lying beneath Błędów was a mineral resource worth exploiting.

Early medieval miners could exploit the layers of galena-ore-bearing dolomite limestone exposed in a handful of rock outcrops around the desert. Lead objects found in the region confirm that the local population was mining lead ore in the first millennium, and the 1136 bull of Pope Innocent II makes the first mention of 'silver diggers' in the region. By 1350 the small town of Olkusz near Błędów was flourishing as a centre of lead production, then focused upon underground seams in strata above the water table. After depletion of the most accessible deposits nearest the surface, dewatering was necessary to enable continuing exploitation at deeper levels. Water had to be drawn out of the mines constantly, and was removed mechanically from the mines by hoists and raised to the surface using buckets, ropes and winches and subsequently treadmills operated by pairs of horses – and sometimes humans too. Drainage became the key technical issue determining the viable depth of mining operations.

But by the seventeenth century, these crude and laborious drainage methods proved insufficient to enable the exploitation of yet deeper seams. Long horizontal or nearly horizontal tunnels – drainage adits – were dug underground to discharge water to surface water courses, a much more efficient though expensive means of increasing the depth of exploitation, though with the possible additional benefit of discovering previously unknown ore bodies. Water is typically the element of the natural environment most affected by mining, and dewatering at Błędów caused the water table to drop from 5 m to 30 m across the area, with consequent impacts upon water availability in the plant rooting zone at the surface. The surface water network was affected too, with the nearby Biała Przemsza river and its tributaries contracting in length as water volumes dwindled. In the nineteenth century, steam pumps were installed for more effective dewatering, making it possible to exploit ore deposits at greater depths than adits could achieve.

Intensive development of galena ore mining employing modern techniques commenced in the mid-twentieth century. Dewatering of new and reopened mines in the area beneath the Błędów desert was further accelerated, with the result that the water table dropped 100 m, leading to the area dehydrating even more. When groundwater is abstracted from a site at a rate greater than it is naturally replenished, the result is an underground inverted cone of depression around the well. Depleted of groundwater, the extent of a cone of depression may extend many kilometres around the pumped well, as is the case in the Błędów district where the hydraulic depression cone came to extend over an area of 35,000 ha by the early twenty-first century. Centuries of intensive mining operations had lowered the water table to such an extent that, at the desert surface, plant growth will probably have been restricted.

What had once been a densely wooded landscape had changed completely since early medieval times. The smelting process for lead and silver ores required high temperatures, meaning that from the thirteenth century wood for charcoal was a critical resource for sustaining the mining industry. The wood for charcoal production was supplied through intensive logging of the forests around Błędów, which were subject to repeated phases of deforestation and reforestation for subsequent centuries as mining fortunes waxed and waned. Large-scale deforestation laid waste to the landscape, for once the trees were felled and their roots had died the fragile soil structure disintegrated, once again exposing great expanses of flyaway sand to the whims of the winds and reactivating aeolian processes. In the time of Casimir the Great, King of Poland from 1333 to 1370, huge tracts of shifting sand threatened the very walls of Olkusz. Wood was also needed for constructing underground mining galleries and adits, and a wide range of other uses. So much timber was consumed in support of the extensive mining operations in the Błędów area that it was said whole forests disappeared in mines. Requirements specified by clerks for the nearby Ponikowska adit in 1563 indicated the quantity of oak required for a 100 m adit: 600 pieces for supports, 3,000 for passages, 120 for covers and separators and 300 for wooden shingles, along with timber for wheels, axles, wheelbarrows and shaft-covering roofs. By the end of the sixteenth century, old adit support timbers were being reused; new wood was already scarce at that time. Once the local forests had been comprehensively decimated, wood had to be imported into the region. Clear-felled areas were put to use for livestock grazing rather than being replanted, but suppressing regrowth in this way had the effect of accelerating the destruction of the fragile topsoil.

VAST SHIFTING SANDS

By the early nineteenth century, the area was a desert landscape of vast deflation fields of drifting bleached sand, part of a huge complex of open sands and regenerating woodland which once reached as far as Szczakowa, 25 km to the south-west. Driven by the wind, sand drifts regularly swamped nearby arable land and villages. Olkusz was described by Stanisław Staszic in his 1815 geological account of Poland, *On the Land of the Carpathians and Other Polish Mountains and Plains*, as being surrounded by a sea of impenetrable sands. Forced-labour parties were assembled to plant pines on the dunes surrounding the town, a measure justified on the flimsy grounds that it served as punishment for peasants supposedly having stolen wood from local forests. In an 1841 account of mining in Poland, Łabęcki commented: 'It is difficult on Polish soil to find a [location] less fertile and more like a desert than the location of Olkusz.'[1] The site's vernacular name, 'The Polish Sahara', was coined in 1887 by Wacław Nałkowski, a Polish geographer. The stark differences in appearance and land use compared to its surroundings led Błędów to be characterised as a catastrophe, a place that has experienced devastation, and a medieval ecological disaster. That such places were shunned and scorned as being worthless, ungodly, profane and even hazardous is testified to by the name of another of Poland's numerous European deserts: Diabelskie Pustacie, a nature reserve of nearly 1,000 ha in the north-west of the country, which translates as 'The Devil's Wastelands' or 'Devil's Hollows'. But, haunt of the devil or not, twenty-first-century environmental values offer an alternative perspective on the natural rather than spiritual value of these once-disdained places.

Inevitably the range of species on the desert is limited – for where there's water there's life, and vice versa – but those that are able to survive here are largely restricted to this habitat type, and found in few other Polish locations. Characteristic fauna of this Polish desert include mammals, birds, reptiles and insects alongside Grey Hair-grass and a rich lichen flora common on such dehydrated sites. The Grey Wolf, Eurasian Lynx and Wild Boar are known to be present in the region. An interesting avifauna includes such characteristic desert-steppe species as Red-backed Shrike, Woodlark, Tawny Pipit and Ortolan Bunting. Reptiles include Grass Snake, Sand Lizard and Slow Worm, and amphibians may be encountered on the desert fringes.

Here too the Large Blue butterfly lives out its bizarre life cycle as a brood parasite on an ant host. The larva of the Large Blue spends most of the year within the nest of a single species of red ant, where it merrily feeds upon ant grubs. After pupation, the newly emerged adult butterfly has to clamber out of the nest before taking wing around July, when this handsome mottled sky-blue creature mates and lays its eggs on Wild Thyme. Young larvae burrow into the flower head to feed on the flowers and developing seeds before, once grown to around 4 mm in length, they drop to the ground – where they wait to be found by foraging ants attracted by sweet secretions produced from a special 'honey' gland. Once picked up by an ant and conveyed below ground, the brood parasitic life cycle recommences. Protected by law in Poland, the Large Blue is threatened

---

1 Łabęcki, H. (1841) Wiadomość bibliograficzna o górnictwie w Polsce i naukach przyrodniczych ścisły związek z nim mających. *Biblioteka Warszawska*, 1841 t. 4.

The Red-backed Shrike is a carnivorous bird feeding upon large insects, small birds, frogs, rodents and lizards, the corpses of which it impales on thorns or barbed wire serving as a 'larder'. (Antonios Tsaknakis, CC BY-SA 4.0)

with extinction both there and at the global level; fragmentation and loss of its favoured sparsely vegetated dry sand habitat is the key reason for its decline. The Blue-winged Grasshopper is also very much at home on Błędów as a pioneer species on open, dry habitats free of vegetation. Mottled and buff-grey in colour, the grasshopper's natural camouflage is well adapted to life in such sandy places, yet while scarce in Poland the species does not enjoy legal protection. Another insect typical of such environments is

Large Blue butterfly. (PJC&Co, CC BY-SA 3.0)

the antlion, actually a group of predatory insects which dig inverted cone-shaped pits in sand to trap ants and other small insects. Sporting a powerful pair of sickle-like jaws, the larvae seize their prey and inject venom to immobilise them, and enzymes to digest their soft tissues. Above the sands, dragonflies hawk insects on the wing.

THE POLISH COCHINEAL

In a now long-forgotten episode, a desert insect species with a curious history is the Polish Cochineal. Today found occasionally on the grassed areas of Błędów, this oval-shaped species colloquially known as 'Saint John's blood' is termed a dye animal and was one of few sources of crimson dye for fabrics in sixteenth-century Europe, though its use can be traced back to the late La Tène culture of *c.*125 BCE. Also known as the Polish woodworm, the cochineal is a predominantly sessile parasite on the roots of its host plant, Perennial Knawel, an unexciting-looking herb growing on sandy, dry and infertile soils. A labour-intensive harvesting process entailed uprooting the host plant and painstakingly picking out the female larvae. Each plant yielded about ten insects, each about 5 mm in length. The larvae were killed with boiling water or vinegar and dried in the sun or an oven before the red carminic acid prize was extracted for use in dyeing fabrics. Magnates and noblemen wore robes dyed with Polish Cochineal products, and the resultant bright red colour did not fade over time. The numbers required are sobering: as carminic acid constitutes only 0.6% of the insect's dried body weight, an awful lot of bugs were needed for dyeing processes requiring about 180–250 g of dye per kg of silk, or 50 g per kg of wool. To obtain a single kilogram of red dye, at least 260,000 plants had to be harvested. Whole villages were engaged in the activity, and knawel plantations were established to provide sufficient host plants. This unprepossessing insect was of great economic importance in the Middle Ages, and cochineal became one of Poland's chief exports. But the introduction to Europe in the sixteenth century of cheaper and higher-yielding Mexican cochineal – from a completely different species – and the later development of synthetic dyes ended the economic activity, although the rural Polish continued to harvest the insect for dyeing fabric and as a vodka colourant. With economic redundancy and the dwindling of its habitat, however, the once-widespread Polish Cochineal is now a rare species.

Like the fauna, the diversity of plant life in the Polish desert is relatively restricted. Stanisław Wilecki, a resident of nearby Klucze, recalled that in 1934, when he was 14, 'there was not a single grass in the desert'.[2] Plants there today include Dwarf Everlast, an aromatic perennial herb with woolly leaves growing to 0.3 m in dry soil and flower heads of bright golden-yellow florets, traditionally used in Europe as a medicine. This is another species threatened by habitat loss, as is the Sand Plantain, a flowering plant favouring dry inland sandy areas and on the Polish protected and endangered list. Desert flora also include Bearberry, a drought-tolerant low-lying evergreen shrub of the heath family. Its flowers provide a nectar source for insects, but much more interestingly its

---

2 Dziechciarz, O. (2016, August 12) Błędowska Desert. *Olkuski Przeglad*. https://przeglad.olkuski.pl/pustynia-bledowska/

PUSTYNIA BŁĘDOWSKA: THE POLISH SAHARA

Blue-winged Grasshopper. (Amada44, CC BY 3.0)

The drought-tolerant Perennial Knawel is the host plant of the Polish woodworm insect from which crimson fabric dye was extracted in sixteenth-century Europe. (Krzysztof Ziarnek, Kenraiz, CC BY-SA 4.0)

Bearberry fruiting. (Sten Porse, CC BY-SA 3.0)

Eurasian Brown Bear female. (Charles J. Sharp, CC BY-SA 4.0)

Polish Scurvy-grass is a southern Poland endemic and one of the scarcest and most endangered species of the European flora. (Krzysztof Ziarnek, Kenraiz, CC BY-SA 4.0)

Sand Plantain is an annual herb with a taproot – a species of arid places. (Le.Loup.Gris, CC BY-SA 3.0)

small red fruits are a favourite of the Eurasian Brown Bear. Bear have been recorded in the area for many years, transboundary in habit and probably members of Europe's largest bear population, that of the Carpathian Mountains to the south and east. With a range stretching through Slovakia, Ukraine, Serbia and Romania, forests are the bears' stronghold, although the population, while perhaps numbering 7,000 individuals, is nonetheless classified as vulnerable and in some regions locally endangered. Happily, bears in Europe do not consider humans as potential prey, generally avoiding close encounters.

BŁĘDÓW ENVIRONMENTAL CONDITIONS

The climate of a locality acts as a first-order constraint upon the potential structure of any ecosystem, with long-term mean temperatures being one of the most significant variables; and, perhaps counterintuitively, Błędów can get hot. Despite being routinely referred to as a desert – *pustynia* in Polish – Błędów is subject to Poland's humid continental climate regime, with relatively warm summers and no dry season. But the climate of Poland is transitional between maritime and continental types, and closer examination suggests that a few areas of the country are arguably more accurately classified as dry subhumid. Such climatic zones are more characteristic of the Eurasian steppe, the vast, arid and largely flat belt of treeless grassy plains stretching 8,000 km through Central Asia o Manchuria in the east, whose climate is similar to arid zones in Sub-Saharan Africa and North America.

Broad-scale climatic zones tell some of the story at Błędów, but micro-scale conditions are also significant, for it is at the micro-scale near the ground that ecology happens. Solar radiation is a major determinant of climatic stress, and for many organisms the temperature at the soil surface is the most important thermal feature of the environment. In the hot deserts of the world, extreme temperatures at the soil surface have profound implications for ecology, and while European temperatures are lower the nature of the drivers, processes and impacts are similar on the subcontinent's deserts. Mineral sand retains vanishingly little moisture within the soil or in the air above it, contributing to making Błędów and others more arid than their surroundings. The westerly and south-westerly winds which prevail here have, in tandem with human activities, historically played a key geomorphological role in creating and perpetuating the great deflation surface of the Błędów desert, and its largely unvegetated nature lacking closed-canopy vegetation cover gives desiccating winds free rein. With little ambient moisture, low humidity and consequently a lower heat capacity, Błędów heats relatively rapidly during daylight hours compared to the more verdant lands around the desert, which are more humid as they retain more ambient water and therefore have a higher heat capacity. Temperatures in these surrounding areas increase at a slower rate over the course of the day, for relatively more solar energy is required to heat the air and soil there. While the flora and fauna of central Europe are mostly adapted to the prevailing cool and moist climate, the surface environment of Błędów can be a remarkably hostile one for organism survival and growth.

High ambient temperatures are mostly unknown in the Kraków region, where yearly averages hover around 13 °C and a hot July day might attain 25 °C or more. But at ground level the Błędów desert frequently experiences inordinately high summer temperatures – Stanisław Wilecki, quoted earlier, remarked: 'On a sunny day you couldn't go far with your bare feet, it was so hot.'[3] Such open sandy surfaces constitute an intensive thermal environment, with a relatively low capacity to absorb energy from sunlight and retain heat. Under clear skies, when the soil surface is exposed to direct solar radiation, much of the sun's energy is immediately reflected upwards rather than being absorbed. At about 35–40%, bare sand has a high albedo – the proportion of solar radiation that a surface reflects back into the atmosphere – compared to grassed surfaces, which with a lower albedo of 20–25% absorb more energy, to the extent that only half as much might be reflected upwards compared to bare sand (and forest reflects as little as half as much again). Temperatures on the desert rise rapidly at midsummer, and are highest in the microclimate of the soil–atmosphere interface. Layers of air immediately above the surface are superheated, causing dramatically raised temperatures which may be twice as hot as those half a metre above the surface. Such elevated temperatures immediately above the ground surface are readily perceptible here even on moderately warm days.

While the absence of a weather station on Błędów means that continuous temperature data are not recorded, the few available individual observations are startling. A military habitat conservation team were working on the desert one hot day when high ambient temperatures of about 35 °C caused some of their machinery to malfunction. A May 2000 scientific study reported that, on a dry sunny day with an ambient temperature of 28.7 °C, surface temperatures attained 48.1 °C. Such roasting temperatures might be expected in the world's hot subtropical deserts, but not in temperate central Europe. An interpretation board on the Błędów desert states that the sand heats up to 60 °C in summer, though the veracity of this extraordinary figure is uncertain. In deserts all around the world, heat has significant effects upon ecology. Relatively high temperatures in the soil–atmosphere microclimate exert climate stress upon organisms there, disrupting the life processes of plants including the simplest algae and cyanobacteria, commonly present as pioneer species forming a crust on desertic soils. Heat-stressed plants exhibit low germination rates, growth retardation, reduced photosynthesis and heightened mortality rates. Soil surface temperatures also drop dramatically after sunset, even in summer, and cold stress can prove just as lethal as heat stress, causing cell membranes to rupture, leading to cell death and tissue damage.

Alongside uncommonly high temperatures, these open sandy surfaces also constitute intensive radiative environments, where prolonged exposure to intense ultraviolet light has implications for ecology and organisms. Along with visible light, solar radiation comprises infrared radiation and an ultraviolet-B radiative component. UV-B is readily absorbed by some biomolecules, causing negative impacts at the molecular level – disrupting basic metabolic processes and cellular structures, DNA, proteins and other molecules, and inhibiting growth. While land under grass typically reflects only 2–4% of UV-B, bare sand is a shiny and reflective substrate able to reflect up to 25% of incident

---

3   Dziechciarz, O. (12 August 2016) Błędowska Desert. *Olkuski Przegląd*. https://przeglad.olkuski.pl/pustynia-bledowska/

UV-B radiation, heightening the exposure of plants and animals to UV-B. Though not apparent to the observer, damage to DNA can induce higher mutation rates and increased cell mortality. Moss spores, for instance, common in the desert surface crust, are highly susceptible to UV-B irradiation and may fail to germinate as a result of exposure. UV-B also impairs photosynthesis in many species; reduces size and productivity; and increases plants' susceptibility to disease in general.

Water is another resource critical for plant growth. Sand retains water poorly if at all, and high temperatures exacerbate aridity levels. As the surface heats, moisture is drawn into the atmosphere, a natural effect which is multiplied because the water-retaining capacity of air increases exponentially as temperatures rise. As desiccation progresses, plants have serious problems maintaining their water balance in dehydrated soil, and thus their life processes are impeded, with the lack of water acting as a limiting factor on productivity. Water availability may limit desert ecosystems to a greater extent even than temperatures and rates of radiation. Over such open landscapes the wind plays a significant role, too, being largely unchecked and multiplying the risk of desiccation.

So, the environment of Błędów is hostile to plant establishment, growth and reproduction. In addition to the desperately infertile nature of the growing medium, relatively extreme thermal, radiative and moisture conditions are highly stressful for all kinds of organisms. Those flora best able to survive are sand-loving species – known as psammophytes – adapted to grow in arid soils. Unlike most, such extremophile plants thrive in unshaded areas with plenty of sunlight on infertile and often unstable ground, factors which, acting together, preclude other plants from growing. The presence of psammophytes here is interesting simply because they seem out of place: central Europe is not normally known for a high-temperature environment.

Spotted Knapweed is a pioneer species on habitats with dry acidic soils.
(J.Claude, CC BY-SA 3.0)

Gray Hair-grass is a psammophyte, a plant that grows only in open conditions on sandy soil, typically on beaches and coastal dunes. (Peter Gabler, CC0, via Wikimedia Commons)

Relatively high temperatures are not the only physical phenomena more typically associated with the world's subtropical deserts to be encountered at Błędów, for mirages, sandstorms and fulgurites may also be witnessed here. Mirages are optical phenomena produced by the atmospheric refraction of light rays. They regularly occur over Błędów on hot and calm days: as the sun heats the ground, the hot sand heats the air above. This results in steep thermal gradients between layers at different densities which refract light rays, seeming to bend them – with the effect that images of distant objects or the sky are displaced, producing mirages. Tales of Błędów mirages go back 150 years or more. In one, the Polish botanist and university professor Kazimierz Piech described two separate mirage incidents on Błędów in 1924. In the first, a 'wavy, dark-steel lake surface appeared to us, in which dunes and trees growing [there] were reflected so that the illusion of the existence of a nearby lake was complete'.[4] He and his party of academic visitors observed the phenomenon persisting for over an hour. It seemed so real that his companions could not believe that it was merely a mirage, some stating firmly that there must be a distant lake in the desert. However, having observed numerous similar mirages in the steppe of Astrakhan in southern Russia, the good professor knew exactly what they

4   Piech, K. (1924) Miraże w Pustyni Błędowskiej. *Kosmos* 49: 876–878.

were witnessing: the lake was no more than a displaced image of the sky above, like those frequently observed hovering above hot highways in the summer. Piech observed a second mirage there, less intense than the first, later in the same year. These classic desert phenomena continue to occur at Błędów, with troops reporting having seen mirages while parachuting onto the northern half of the desert.

Another extraordinary optical phenomenon here, though less common than the mirage, is looming. In 1863, a Captain Maciej Konecki described an incident when, as a detachment of Garibaldian soldiers marched through the Błędów desert, suddenly on the north-eastern horizon they saw an image of people marching along a road through a village. Konecki recognised it to be a distant image of Russian troops moving west and immediately informed his superiors, with the effect that, two hours later, the Russians were intercepted. If this story can be believed, it resembles an instance of looming, an unlikely but real optical atmospheric refraction phenomenon whereby distant objects and landscapes located beyond the horizon are visible above it. Looming is caused by abnormally powerful refraction, usually due to a temperature inversion caused when layers of air near the Earth's surface are colder than the air above it. Rays of light from objects below the horizon that would normally pass over the observer's head are refracted downwards, thus making distant objects and landscapes visible. Another account related by Mieczysław Prom-Fiołek tells how, one day in 1942 after a long and exhausting journey, he was forced to spend the night in an isolated and dilapidated cottage on Błędów. His hosts – described as decrepit, mouldy and miserable old men living in poverty – related tales of when 'ominous signs in the sky'[5] appeared over the desert on occasion, which were interpreted as heralding war, crop failures, disease and other misfortunes. Intrigued by their fantastical tales, Mieczysław ventured into the desert on one hot August day and suddenly, on the horizon, images of distant landscapes began to form: images of forests, fields and cottages scattered in clearings.

Sandstorms are not at all common in central Europe, but these sometimes violent phenomena occur at Błędów. When in arid areas the ground is dry, and hot sun results in sufficiently high soil surface temperatures, vigorous updraughts spiral upwards from the ground. Hot air rising in the form of a narrow column whirling around a roughly vertical axis sucks dust and sand into the air to heights from a few metres to sometimes over thirty. Particles are spread widely over the surrounding area and the dust devil may even travel along the ground until, after a few minutes, fresh air sucked into the base of the vortex cools the ground and cuts off its heat supply. More conventional sandstorms here are caused by strong winds lofting sand and dust from dry areas across the length and breadth of Błędów and conveying them downwind before deposition, over often large distances. During sandstorms in 2022, traversing the desert was challenging because visibility was reduced to almost zero. Visitor information boards on Błędów affirm that sandstorms are potentially dangerous, counselling that visitors should exercise caution on stormy days. The desert's exceptional summertime temperatures can also be hazardous

---

5   Rzeczycki, T. (27 March 2013) Legenda pustyni nieprawdziwej (cz. 1). *Interia Wydarzenia.* https://wydarzenia.interia.pl/historia/news-legenda-pustyni-nieprawdziwej-cz-1,nId,947088

for humans, for on the hottest summer days the sand burns and visitors are forced to seek out shade: advice displayed there directs that children wear appropriate headgear and carry plenty of drinking water.

Literally lightning rocks, fulgurites are another natural phenomenon associated with the world's deserts to be found on Błedów. When lightning strikes, magic happens: as the electrical charge blasts through the sand, a bolt of 100 million volts may generate temperatures of 2,500 °C or more. At about 1800 °C sand melts, and underground the quartz mineral grains are instantly fused and vitrified, producing structures resembling forking tree roots. These contorted glass-like tubes a few centimetres in diameter are hollow and smooth inside, with a rough outside surface covered with partially melted sand grains, and while generally less than 3 m in length they may extend to 15 m. Although uncommon, this novel petrified-lightning geological treasure may be picked from the desert surface and used to create jewellery. The presence of fulgurites offers a reminder that extreme care needs to be taken when an electrical storm is in prospect at Błedów and similarly exposed desertic sites, where convectional updraughts make lightning strikes a regular feature.

MILITARY RANGE

The military recognised the potential of this neglected and seemingly redundant central European wasteland, although not for its biodiversity value. Błedów was used and abused for military training and weapons proving throughout the twentieth century. During the First World War, it was employed as a training ground and shooting range for Polish troops preparing to fight the Russians in a famous 1914 victory at nearby Krzywopłoty. In the interwar period, the Polish desert was used for military aviation training, with targets including mock-ups of planes and wrecked tank hulls positioned on the southern half for aircraft shooting practice, though in 1939 a PZL.23 Karaś light bomber crashed onto the Błedów desert killing its crew of three. In the later 1930s, infantry, cavalry and artillery manoeuvres were held here.

During the Second World War, the desert was used intensively by the occupying Nazis for troop training. Błedów's resemblance to the deserts of North Africa made it an ideal training environment for the troops of the Afrika Korps, the Nazi German expeditionary force in Africa during the North African campaign. It is uncertain whether Field Marshal Erwin Rommel visited Błedów, but had he done so he would have watched Panzer tanks and other military vehicles negotiating the parched sands before deployment to the Sahara. The remains of bunkers built at this time can still be seen around the margins. The Błedów desert was also employed as a weapons proving ground for the Luftwaffe, where new types of bombs weighing up to 1,000 kg were dropped from aircraft, as well as incendiary and cluster bombs and radar countermeasure devices that released thousands of strips of aluminium foil. V1 flying bombs and V2 rocket missiles are thought to have been tested here too. Błedów was subjected to a hellish firestorm of explosive abuse. Local residents were forced into labour parties to undertake maintenance tasks on the military range.

After the Second World War, Błędów was once again taken over by the Polish military. An area of 2,300 ha was expropriated for use as an aviation training ground, including for MiG-17 and MiG-21 jet fighter aircraft, and military vehicle exercises. No scientific investigations were permitted. Such intensive and prolonged military use may actually have contributed to maintaining the desertic character of the area, preserving open sand surfaces as tanks ripped up the soil, shells, bombs and missiles exploded, woods caught fire, trenches were dug and trampling knocked the invasive vegetation back. But a century of military use left a deadly legacy of explosives that had failed to detonate buried beneath the seemingly innocuous sands. Inevitably perhaps, there were numerous accidents, the worst of which was in March 1979 when four boys from a local school were killed when a bomb they were examining exploded; after this, live munitions were no longer used.

In addition to military abuse, for decades the site suffered the further indignity of being used as a casual garbage dump. Błędów must have presented a sorry sight as the desert became a post-war scrapyard. Until the 1970s, wrecks of tanks, anti-aircraft guns and other military hardware were dumped here before finally being transported to steel mills to be recycled. Some were dismantled for their own uses by enterprising local people, who created everyday items from formerly live shells and artillery components were repurposed as agricultural equipment and used in sawmills. Fires were frequent. Also, in the 1950s, liquid industrial waste was piped into the Biała Przemsza river, causing significant pollution of surface and underground water. The first environmental survey conducted on this hell's graveyard in the late 1980s found an alarming range of environmental contaminants which had to be cleared before conservation work and public access could proceed safely. Hazardous materials included mercury, lead and sulphur, alongside the more conventional but similarly undesirable garbage which had accumulated over decades of casual fly-tipping. Beginning in 2012, a large-scale project detected and cleared thousands of items over an area of 400 ha. Nearly 4,000 dangerous objects were located, including hundreds of terrifying-looking undetonated bombs and shells, many of which were blown up onsite in specially prepared pits. The armed forces' operational area reduced in size, but the northern half still accommodates a military training ground of 375 ha where Polish paratroops regularly train. In 1999, NATO staged military manoeuvres here involving troops from 18 countries, and again in 2014 when Polish, Canadian and American troops jumped onto the Błędów desert along with military vehicles and equipment. These mass airborne troop drops have proved a popular spectacle with residents and visitors alike.

AFFORESTATION

In this dynamic landscape, the area of open desert has expanded and contracted over millennia, influenced by natural succession, human intervention and climate change. But one constant throughout the centuries has been the threat of flyaway drift sand, which continued to cause problems for the local community well into the twentieth century. Local householders who had no choice but to regularly sweep their porches, gardens and driveways, and farmers resigned to seeing their fields swamped after storms, resolved to take action. From mid-century, large-scale tree planting commenced on the western

Sharp-leaved Willow. (Panek, CC BY-SA 4.0)

part of the desert. While the deficiencies of the growing medium frequently resulted in undersized and poorly developed trees, afforestation acted as a windbreak, sheltering the soil surface and bringing shade and humidity to the desert. Large expanses of bare sand rapidly turned green, mitigating the nuisance and frustration caused by unruly drifting sand. Pine and Silver Birch were planted, and also an alien species: Sharp-leaved Willow was introduced on the desert. This large deciduous shrub native to Russia and Central Asia is an early coloniser, commonly planted to stabilise coastal dunes. It is wind-resistant, frost-hardy, withstands sand-blow and promotes the formation of soil in sandy areas. Sharp-leaved Willow spread like lava across Błędów.

Afforestation of the desert continued until the 1990s. Unfortunately, however, while successfully reducing the area of flyaway sand, the tree planting programme was wholly inimical to the site's biodiversity and landscape interest. For some, Sharp-leaved Willow was a saviour shrub; for others, it had become an invasive menace, acting as a nurse crop paving the way for the establishment of pine and other vegetation. Vital aeolian processes were suppressed, and as organic matter accumulated on Błędów, soil profiles began developing. Aerial images show that the extent of bare sand declined from 20% in 1974 to 3.2% in 1999, and to a paltry 1.1% by 2010.

At the start of the twenty-first century, however, recognition was dawning that the local community was losing part of its identity, part of its own environmental heritage.

Also, as a prize example of an increasingly scarce European dry terrestrial ecosystem, the desert was a site of importance at the European level. Błędów was once the largest expanse of drift sand in Poland; one of the few in central Europe where active aeolian processes continued on a large scale; and an increasingly rare instance of a location where sand hill surface geology is visible rather than obscured by vegetation. Its ecosystem processes are recognisably natural, providing home range for numerous scarce and legally protected species of fauna and flora. A remnant pimple of wilderness, Błędów was a declining enclave of biodiversity with a distinctive and arguably beautiful landscape set within an urbanised region. The site also served as a reminder of southern Poland's socio-economic and cultural heritage, and its otherworldly atmosphere and role as a stage for mirages and other exotic desertic phenomena confirmed it as an extraordinary and exceptional environment. Błędów was worthy of conservation. It needed to be saved for the future.

NATURE RESERVE DESIGNATION

Błędów was designated as a nature reserve of European significance in 2008 for its inland dune and dry grassland habitats. Substantial funding was obtained for two European Union LIFE conservation projects over the period 2011–17. In a large-scale and lengthy operation, the desert was systematically stripped of vegetation. Trees, scrub and shrubs shading the soil were cleared wholesale, roots grubbed up and surface litter removed and burned across an area of 375 ha, restoring a surface of bare sand. Physical features restricting wind activity were removed with the aim of restarting the wind-driven geomorphological system, to re-establish desertic processes. Thus with a huge effort the site was restored to a favourable conservation status, though it is a little ironic that, to get there, Błędów was yet again deforested, but this time in a planned and carefully managed fashion, and with motivations very different to the hard-nosed economic ones that drove historic deforestation episodes here.

The extensive habitat management work at Błędów served more than just conservation goals. With one eye on growing markets for ecotourism, local authorities had recognised the tourist potential of the desert on their doorstep. Granted, the deserts of Europe are not usually major tourist attractions, and few have been much developed for tourism and recreation. But so singular is it that this desert is renowned throughout Poland as Pustynia Błędowska, and a new visitor facility on the southern edge of the desert opened in 2018 provides a focus for visitor access. Now a star-shaped land pier extends 40 m over the sands, information boards interpret the desert for the non-specialist, and a Desert Information Centre has been established in the nearby town of Klucze together with a web page. Designated trails for students, walkers and horse and motocross riders make this for some forbidding wide-open space more approachable and accessible, and most are confined to the desert's eastern extreme to minimise disturbance. Paragliding is a popular activity here, and on hot days the casual visitor may enjoy the sight of brave aviators skilfully exploiting rising columns of hot air – thermals – as sources of lift as they soar overhead. While conservation remains the primary goal on this most famous Polish desert, developing tourism and recreation offers a potentially new economic role: monetising the resource, on however limited a scale. Now Błędów is a European desert with the visitor infrastructure necessary to enable the public to appreciate the qualities of the place, rather than remaining an unknown and uncherished backwater like so many others.

Visitor developments will perhaps also serve to build support for conservation goals at Błędów. Both Nowa Huta and the Olkusz lead-silver mine have ceased operating, so maybe the desert has a twenty-first-century role as a symbol of changing values amidst the deindustrialisation of this former heavy-industry region. A site until relatively recently attributed a low or zero value, and seen as fit only for depreciative military use and waste disposal, now has a long-term conservation management plan. Nature never sleeps, however, meaning that habitat management work involving clearance of invasive vegetation will need to be maintained long-term to sustain the biodiversity and landscape interest. Tasks include maintaining and expanding active deflation fields and the area of grassland habitat, environmental monitoring, ecological education and visitor management measures. Such ambitious plans do not come cheap: the annual recurring cost was estimated in 2017 at nearly €200,000. So a little more irony: a lot of money is being devoted to preserve a 'medieval ecological disaster'! But perhaps at Błędów we can discern a way forward for many of the subcontinent's neglected deserts, and particularly the numerous other sites with histories as military training grounds.

Tearing myself away from this paragon of European deserts was not easy. Hot, parched, barren, subject to mirages and sandstorms, and a locus for fulgurite rocks forged in the heat of terrifying lightning strikes, the thrilling sand sea that is Błędów lives again. No more mining, no more afforestation, and no more of the grotesque depredations inflicted by military use. Desert-loving species once again have free rein. The relentlessly barren Błędów landscape is like nowhere else. All this resumed activity on the desert, however, clearly risks sparking the resumption of drift-sand migration, the old enemy hereabouts. Maintaining the local community's identity and environmental heritage is a laudable aim, but the question remains: in perpetuating one of the largest tracts of drift sands in central Europe, will Błędów's flighty sand once again be unleashed to infuriate householders and farmers, in the same way it has for the last 2,000 years or more?

Chapter Six

ICELAND: COLD BLACK DESERTS

Probably the best-known desert islands in the North Atlantic Ocean are the Canary Islands. Lying 100 km off Morocco, the Spanish archipelago of seven main islands enjoys a warm subtropical climate. La Palma and Tenerife islands are particularly popular with north European tourists seeking beaches, unspoiled nature and highly developed modern resorts, especially in autumn and spring. Arid conditions predominate, and water is scarce on these rocky and volcanic specks. They capture some of the wind-transported Saharan mineral dust swept from the world's largest hot subtropical desert, plumes so dense on occasion that airports have been forced to close. One of the most popular Islas Canarias, Fuerteventura, is commonly described as a desert island, where an arid and salty terrestrial environment militates against plant life thriving.

Iceland is another desert island in the North Atlantic, though this one is only rarely recognised as such. However unlikely it may seem, here is the largest desert in Europe. The flight into Keflavík airport brings into view a large island with only a few scant areas of green. Aerial views were largely obscured by brooding low clouds hanging over the country, and with news of a volcanic eruption taking place at Litli-Hrútur, barely 40 km from the capital Reykjavík, the traveller's mind inevitably recalls uncomfortable memories of the time when the Eyjafjallajökull volcano famously erupted for 39 days, spreading a plume of fine ash over large parts of Europe and the North Atlantic. Volcanic activity brought unprecedented and widespread disruption to air travel over Europe for several days, and although relatively modest in comparison to some historic Icelandic eruptions, Eyjafjöll caused over 100,000 flight cancellations affecting seven million passengers, reaffirming the dominance of the planet's physical environment over the human one. Driving out of prim Reykjavík, the world's northernmost capital city set in a landscape of rocky massifs, it quickly became clear that Iceland (*Ísland*) is a country dominated by natural forces, principally volcanism – the most powerful and most terrible of the planetary processes – but also a more familiar one: the wind. Both contribute to making Iceland a unique instance of a European desert, and a seldom-seen young and actively erosional landscape on a grand scale.

With a wholly European history and culture, Iceland is conventionally assigned as European by reasons of geography, history, culture and language, though the country has been described unceremoniously as a 'marginal society' by Gunnar Karlsson, formerly Professor of History at the University of Iceland. It's a big place: huge, really, Europe's second-largest island after Great Britain, but with a tiny population of 387,758, about the same as Cardiff. It is one of the world's most urban-concentrated major countries, its inhabitants mostly living in a few locations hugging the west, north and eastern coasts. Most of the island is barren. The map of the country is vanishingly sparse: no tarmacked roads cross the empty interior, most mountain roads are closed until the end of June because of snow and muddy conditions making them impassable, and there is no railway. Located at a point where the North American and Eurasian tectonic plates meet, this desert island is remote by any standards, its nearest neighbour being Greenland 290 km away and the centre of green Europe lying about 3,500 km distant.

DESERT ISLAND

Iceland is an extraordinary place: a desert island with parallel. Physically, the country comprises a huge block, a high and mountainous plateau fashioned by eruptive processes, shaped by erosive forces, and surrounded by coastal lowlands and fjords. A geologically youthful place, it is consequently rough-hewn, abrupt and jagged, frequently lacking the softness of outline characterising more mature landscapes. The area of the interior known as the Highland (*not* Highlands) is perhaps the largest area of terrestrial wilderness in Europe, where vast areas of barren sand exist alongside glaciers, volcanoes, hot springs, lava fields and a few scattered oases of vegetation. While unremarkable to the Icelanders, the country's extensive black volcanic deserts are unique in global terms. These are the most barren desertic wastelands in all of Europe, uninhabited and uninhabitable, never cultivated, and exceptional in their extent and extreme character. Nowhere else comes close. To reassert: deserts are not commonly associated with Europe, nor indeed with the planet's most northerly regions, which are more typically under tundra and conifers. Of course, Iceland's deserts differ in several ways from other European deserts, most notably in being cold. Well, yet colder than those on the subcontinental mainland. Iceland is not generally recognised as desertic because the deserts of the popular imagination are a subtropical phenomenon, not one located near the chilly roof of the world. The point is acknowledged, perhaps a little wearily, by Professor Olafur Arnalds, soil scientist at the Agricultural University of Iceland, who asserts: 'These [Icelandic] deserts are seldom reported in surveys of large sandy areas of the world.'[1] A similar observation can be applied to the deserts found throughout Europe.

A full 40% of the island is arid desert. Revelling in dramatic Icelandic names like Dyngjusandur, Skeiðarársandur, Mýrdalssandur and Mælifellssandur, extensive desert areas stretch from coastal sand-fields to remote inland regions, from the north-east through the centre to the south coast of this loosely oval-shaped republic. The composition of their surfaces take one of three forms:

- sandy gravel, coarse pavement surfaces of the granules-to-cobbles-sized residue left once finer-scale particles have been winnowed out by wind or water

---

1 Arnalds, O., Gisladottir, F.O. and Sigurjonsson, H. (2001) Sandy deserts of Iceland: an overview. *Journal of Arid Environments* 47(3): 359–371.

- lava surfaces upon which sand and volcanic ash have been deposited by the wind and eruptions; and
- deserts of fine-grained black sand, locally known as sandur and covering one-fifth of Iceland – the focus of this chapter.

These latter free-draining volcanic glass sandur soils are shallow, coarse and low in organic matter, and their limited ability to support plant growth makes them largely unsuitable for the grazing and other land uses sustained on some other Icelandic soils. To those more accustomed to a verdant Europe, much of Iceland appears relentlessly bleak, with some land-use maps classifying much of the land surface as, variously, 'Waste Land', 'Other Land', 'Open Spaces' and 'Bare Soils'. Only one-fifth is under grazing pasture – Icelandic agriculture is primarily based on dairy production and sheep farming – barely 1% is under arable crops, mostly in the lowlands, about 1.5% is wooded land, and about 10% is permanently smothered under fields of ice.

Getting to experience these black-sand deserts is not straightforward, however, because many are remote and only accessible via dirt tracks, and then a hike. After struggling to get there, the observer is presented with vast and often featureless expanses varying in extent from a scant few to thousands of hectares, with many places being flat all the way to the fuzzy horizon. There's just one colourway, and that's grey-black. Matt. Monotone. Monochromatic. Monotonous. Melancholic. No trees, no shrubs, mostly. Here, the very daylight seems strained; precious. The deserts on the European subcontinent are generally high-intensity sunwashed environments: light, bright and relatively cheerful. But here on Iceland's deserts the daylight seems strained, lacklustre, denatured, subdued, any life relentlessly squeezed out of it like some horrific terrestrial black hole. The wind blasts and roars ceaselessly across this patchily vegetated habitat, though there are few other sound sources here. One senses some sort of innate though vague and unstated threat in the air. Perhaps this is what hell is like. Everything seems imminently in motion, everything feels uneasy, temporary, unfixed, provisional. Gobbets of powdered black volcanic dust are hurled across the road on the slightest breeze. These landscapes are so extraordinarily far from everyday experience it is difficult to identify comparator European environments: in their crushed desolation perhaps only some of the derelict former heavy-industry sites of urban Europe come close. The occasional tarmacked public highways blend well with their gritty surroundings, with only the central white lines and yellow-painted traffic snow posts really standing out in this horrorscape. Otherwise, there are vanishingly few other traces of humanity in prospect. No one lives here. No one lives anywhere near here either, mostly, for the sandy deserts are uniformly bereft of people.

Yet there is a sense of peacefulness and stark beauty – perhaps a terrible beauty – to be had among these outstandingly wild and even romantic landscapes. Also the thrill of the realisation that such an otherworldly environment exists at all in Europe, one which in parts has never been cultivated or inhabited… ever. Nonetheless, somehow subdued and almost depressed, you put your head down, you climb back into your vehicle, and you get out of there.

The Icelandic black-sand wildernesses are similar in many ways to other deserts around the planet. These are proper deserts, arid and bare sandy plains. A key difference from many other European locations, however, is that the sand here is of volcanic origin: the particles are predominantly of basalt with material from volcanic eruptions, rather than the quartz-rich sand sheets more typical of Europe's deserts. This basaltic volcanic glass sand is hard, angular, abrasive and mildly corrosive, with low organic content and very low levels of nitrogen – materials typical of such youthful and transient landscapes. It is the basalt that creates the brooding dark surfaces covering much of the island, simply because its composition is dominated by minerals rich in magnesium and ferric oxides which produce the characteristic grey-black colour. Another difference centres upon the sizes of the individual grains, which are more akin to dust than the rounded quartz grains more common elsewhere, and much more readily blown around as a consequence. Iceland is one of the dustiest places in the world, and with little vegetation and even less soil to shelter the ground surface the fine black sand drifts readily, and in huge quantities. Intense storms cause some of the most extreme wind erosion events recorded on Earth, for when wind velocities attain 20 m/sec – gale force – with violent storm gusts exceeding 35 m/sec, enormous amounts of dry desert sand are entrained across this flat and unobstructed land, and the brave observer in the field can only boggle as airborne sand whips by at a rate of two tons or more per hour at head height. In addition, surface creep may boost this wind erosion figure by 10–30% as larger particles are slid or rolled along the ground and smaller ones are lofted into the air, bouncing across the surface and hurling other particles into the air through 'splashing', thus escalating the process in a phenomenon known as saltation (jumping). While much black sand is deposited on the island, plenty is blown offshore across the southern Iceland coastline, and throughout the year including winter and at sub-zero temperatures.

So otherworldly are the resultant profoundly dark-hued landscapes that geologists – not generally renowned for imaginative characterisations – have likened them to those of Mars, and have studied them to better understand aeolian processes. These European deserts are unique not only as the world's largest volcaniclastic deserts but also for their extraterrestrial appearance! They feature numerous small aeolian features including surface ripples, ventifacts – rocks that have been abraded, pitted, etched, grooved, or polished by wind-driven grains – and, at a few coastal locations, swarms of black and even yellow sand dunes. But, curiously, unlike other European deserts, inland dunes are scarce and mostly low in relief, though they may develop where obstacles in the path of the sand cause accumulation in the lee of the wind. The scarcity of dunes is probably attributable to the island's regime of near-constant and frequently high-velocity winds, which whisk loose surface materials away with great efficiency. Up to 130 so-called dust events occur each year, and with a single storm capable of generating 500,000 tons or more of dust, the effect may be that dunes are simply unable to form. A further factor may be particle shapes: basalt sand here is not rounded like grains of quartz, and is more readily lofted airborne as a result.

GLACIERS

From where does all this flyaway grey-black material originate? The largest sand seas on earth frequently owe their continued existence to a sustained supply of parent material, and Iceland's sources are so active that the country's total emissions rank it among the most active dust sources on Earth. Here, desert sand originates from two main sources: glaciers grinding volcanic rock to a fine powder, and fine sand in the form of ash deposited in volcanic eruptions. A novel phenomenon on the European deserts, a full one-tenth of the country lies under glaciers. A glacier is a large mass of perennial ice formed in areas where the temperature is usually below freezing, and replenished by snow recrystallising and being compressed into solid ice over time, with trapped air being progressively squeezed out. Glaciers play a major role in shaping the landscape as they flow inexorably downslope, propelled by their own weight in addition to gravity. Eroding, weathering, transporting and depositing, glaciers produce, move and modify the large quantities of silt and sand generated as rocks that have been frozen into the base and sides of the glacier mechanically abrade the underlying rock like giant geological sandpaper. This is a contest of unequals: slow but relentless, and the result a forgone conclusion. Rock flour is the product of this relentless grinding process, the potential supply of which is essentially unlimited. One indication of the potential quantities involved is a figure for the giant Greenland ice sheet, which produces one billion tons of glacial flour each year. Fine particles thus generated are transported from underneath the glaciers, flowing as mud in subglacial meltwater rivers which deposit their load at the glacial margins and as outwash deposits on floodplains at the end of the glacier. Yet more is deposited in the terminal moraine, a ridge, mound, or even plain of loose sediments and rock debris laid down or pushed up by a glacier at its foot or snout. Lagoons often form close to glacial margins, and when they dry out – which may occur as frequently as daily – their sandy deposits become yet another source of aeolian materials subsequently transported over large areas.

As glaciers recede, they dump yet more silt and sand. The largest glacier in Iceland is Vatnajökull, a mega-glacier of 810,000 ha and a maximum depth of nearly one kilometre. Vatnajökull is a major source of sand, mostly in the form of deposits at the margins of the ice. Tributary streams issuing from the ice front loaded with glacial sediments flood wide areas, often daily during warm weather, and when the waters recede large quantities of unstable fine sediment are spirited away by the wind, providing a sustained supply of parent material for the deserts. One such glacial margin floodplain is the giant Skeiðarársandur black sand field in southern Iceland – at 90,000 ha nearly double the size of the urban area of Birmingham, UK – which experienced a catastrophic flood after the 1996 volcanic eruption under the Vatnajökull glacier as rapid ice-melt deposited volcanic material over large areas. A phenomenon both scarce and extraordinary at the global scale, such eruptions may reach temperatures of up to 1,200 °C. As the heat melts the ice from below, the meltwater gradually lifts the ice cap before bursting out forcefully from under the glacier, and creating a flood on the outwash plain. Eruptions under glaciers constitute a major source of sand materials in Iceland. Extreme instances of such events can produce discharge rates of 100,000 to more than 200,000 cubic metres per second, comparable to some of the largest rivers on Earth. Immense areas may be flooded, sweeping away soil wholesale to produce unstable surface glaciofluvial deposits, and creating and recharging active aeolian areas from which sandy deserts subsequently advance. Such eruptions under ice are called *jökulhlaups* in Icelandic – literally, glacial outburst floods. The frequent and large-scale occurrence of such dramatic fire and ice spectaculars is one of the earth science properties that characterise Iceland's sandy deserts. Indeed, the magnitude and frequency of *jökulhlaups* are unique to Iceland, and the Icelandic term is employed internationally for the phenomenon.

VOLCANISM AND DUST

The most significant aspect of the geology of Iceland is its volcanism, and eruptions are the second major source of sandy material feeding Iceland's cold deserts. The Iceland hotspot is one of the most volcanically active regions on Earth, with over 200 volcanoes. The country is a volcanic wonderland, sitting as it does over the subsurface Iceland plume, an unusually hot mantle plume acting like a chimney slowly transporting hot, buoyant rock from deep inside the Earth, which generates frequent and extensive volcanic activity and hence dust production, and to which Iceland owes its very existence. Eruptions on a scale deemed minor on the island but that would cause unimaginable terror elsewhere in Europe are gleefully described as 'tourist eruptions'. Eruptions produce desert sand in the form of fine volcanic rock ash and glass: glass because, when molten basalt is flung into the air and cools rapidly, the fragments have no opportunity to recrystallise but instead turn into natural glass. Such deposits are widespread across Iceland, and even moderate-sized eruptions may result in volcanic ash reaching northern Europe and Scandinavia. Around 30–40 million tons of dust is produced annually, ranking Iceland among the most active dust sources on Earth and comparable to the warm deserts of the world. The sheer scale of dust emissions from the Icelandic deserts is thought to be significant at the global scale, perhaps affecting atmospheric conditions in both the North Atlantic Ocean and the Arctic, and has consequential impacts upon every ecosystem on Iceland. With eruptions occurring every three to six years on average, fresh volcanic material is produced on a large scale, resulting in deposition layers with thicknesses ranging from a few millimetres to 10 cm or more. Major eruptions can produce staggering amounts, such as the several cubic kilometres of rock fragments flung into the atmosphere by Mount Hekla, Iceland's best-known and most active volcano, which has erupted more than 20 times since 874. In 1783, the Laki eruption took the form of a volcanic fissure with 130 craters opening in the south of Iceland, which raged for several months and spread material over about 8% of the country's land area. The smallest fragments comprise particles of 2 mm and less in diameter, and are readily lofted in the volcanic plume to several kilometres in height before being transported over great distances by the wind, and though tiny can constitute a major hazard to aviation. The largest fragments are termed blocks and bombs, and pose a significant hazard to humans, having historically caused fatalities. The threat remains: Icelandic volcanologists are all too aware that it is not a matter of if but when the next Laki-like eruption will occur.

A showpiece of volcanic geology, Ódáðahraun in the central Highland is the largest desert in Iceland. This vast uninhabited wilderness of sandy plains and lava surfaces covers an area of 4,400 km^2 or more, a blasted grey-black hellhole looking like nowhere else in Europe. Ódáðahraun puts the desolate in desert, being largely flat, subject to intense sandstorms and occasional volcanic fissure eruptions, and having practically no vegetation. It is also profoundly arid, located in the relatively dry rain shadow cast by the Vatnajökull ice cap, meaning minimal precipitation. Neither are there any streams in this highly permeable area, only at its edges: plucky hikers seeking to journey over the wilderness are advised to carry plenty of water. The Ódáðahraun desert is a major aeolian sand transportation pathway, along which sand deposits from the ice margin of Vatnajökull and volcanic eruptions and flood deposits are whisked off northwards for tens of kilometres. Set at 728 m, this remote and little-known region of Iceland is accessible only with 4×4 vehicles or on foot, and is not a popular destination for the tourist, even sand-boarders – Iceland's take on snowboarding. The largest and most active dust-plume area, Dyngjusandur, is located here. Described as 'hyperactive', Dyngjusandur is one of the most prolific dust sources on Earth, every square metre capable of emitting large quantities of fine basaltic sandy volcanic glass. Intense dust storms on windy days can loft several hundred thousand tons airborne, causing dust pollution capable of obscuring the view for tens of kilometres.

About three-quarters of the 2,200,000 ha of the fine black-sand deserts of Iceland's interior are active aeolian surfaces. Within these, several locations have been identified

Iceland – Ódáðahraun panorama. (Pietro, CC BY-SA 3.0)

Lichen *Stereocaulon vesuvianum* from Hekla in Rangárvallasýsla, Iceland, where it flourishes on lava fields. (Jóhannesbjarki, CC BY-SA 4.0)

and termed *plume areas*, places that generate dust most frequently and where aeolian erosion rates are the highest. The combined result of plate tectonics, a deep-Earth-mantle hotspot, slow but unremitting glacial rock flour production, and intense aeolian weathering processes is an extraordinary desert-forming and -sustaining system unique in Europe. The desert geomorphological system in the remote Icelandic countryside might best be understood as analogous to that of a conveyor-belt production-line system:

- fine sand of volcanic origin is the input for the system;
- desert expanses are continually recharged with fine volcanic sand from eruptions, aeolian deposition and surface transport from various sources;
- more or less permanent deserts are formed, which however are unconsolidated in nature (loosely arranged or unstratified, or whose particles are not cemented together);
- frequent high winds sweep much of this friable and dry material away in deflation processes – wind erosion of dry and loose material in blow-out zones, where sand is continuously removed by aeolian action resulting in the ground surface physically lowering over time;
- soils and vegetation have little opportunity to develop, with the outcome being that the expanses of cold arid desert are perpetuated;
- continual inputs of fine volcanic sand sustain the Icelandic desert formation system.

Desert dust has been identified as an important climate change driver when the deposition of light-absorbing particles upon snow, sea ice, and glaciers darkens their surfaces and accelerates melting through reduction of the albedo – the degree to which they reflect the light and heat of the sun back into the atmosphere rather than absorbing them – of their otherwise highly reflective virgin-white surfaces.

Climate plays a major role in maintaining the Icelandic deserts. Located immediately south of the Arctic Circle, Iceland's climate is relatively mild given its far northerly global position. A humid and predominantly cool temperate maritime climate is substantially moderated by the powerful North Atlantic Drift ocean current which visits relatively warm water to its shores year-round, maintaining the island almost entirely free from the permafrost common over much of the planet's northern margins and without which Iceland's climate would be considerably colder than it is. The Icelandic Low, a large and persistent atmospheric low-pressure system that forms between Iceland and southern Greenland, dominates wind circulation over the North Atlantic and causes high wind velocities on the island. With moderate rainfall and drying winds combined with free-draining surface materials, the Highland interior is described as a cold desert. The weather can and does change rapidly and dramatically, and the adventurer may encounter rain, snow, sun, wind and fog all within the same hour – or even in a matter of a few minutes, hence the common Icelandic saying: 'If you don't like the weather, just wait five minutes.' With as few as four hours of daylight in December and January, and the chill factor derived from persistent strong winds, the winter months certainly feel like the trough of the year. Overall, the combination of the prevailing climate, masses of unstable fine desertic material, and bare open landscapes subject to a regime of constant wind and frequent and intense storms contribute to the maintenance of barely or unvegetated ground surfaces with little or no soil development, and large deflation plains on a scale more normally associated with hot deserts like the Sahara. The erosive power of the prevailing climate is well illustrated by the profiles of mountains set in the black-sand plains of the Highland, which often exhibit a distinct abrasion line on their lower slopes: while perhaps vegetated above, below this line the exposed sides of the hill landform are heavily abraded, blasted by wind-whipped particles acting like sandpaper in the landscape to the extent that a notch may be visible in the mountain's side profile.

ENVIRONMENTAL HISTORY

Iceland has been described by Jared Diamond, the American scientist and biogeographer, as 'ecologically the most heavily damaged country in Europe'.[2] Why, when, and how did this happen? Iceland's deserts are unusual for several reasons, and notable here for the fact that much is known about their history since the island's first settlement. The early history of many other European deserts is largely unknown, for humanity's arrival and subsequent impacts on their environments commenced perhaps five to ten thousand years ago or more, leaving no contemporary written record. In Iceland, however, human

---

2   Diamond, J. (2005) *Collapse: How Societies Choose to Fail Or Succeed.* New York: Viking. https://doi.org/10.62608/2158-0669.1017

settlement and land use have been proceeding for little over 1,100 years, a period which in comparison has been relatively well documented. Unlike elsewhere, then, there are historic records – however imperfect – from the earliest times, which offer insights into the desert island's first settlement and incoming land-use practices, and how interactions between society, economy and the natural environment changed over time. Another noteworthy point of differentiation is that they have been extensively studied, once again unlike many others in Europe.

The story of the Icelandic environment over the last millennium is a sobering one. Some historians maintain that a handful of Irish monks may have been living on this faraway cold-desert island when the first settlers arrived, but Iceland is thought to have been essentially uninhabited at that time (and had there been any Irish monks, Ari Þorgilsson the Learned, the first Icelandic historian, asserts in his *Book of the Icelanders* that they would have left, not wishing to live among heathens). In a relatively stable environment before *landnám* – the Norse settlement – in or around 874, Iceland's lowland areas sustained wetlands, woodlands and heath, while many upland areas supported shrubs and woodland among heath and open areas. Estimates vary, but about two-thirds of the area of Iceland may have been vegetated at the time of *landnám*, with birch woodland covering 25–40% and dwarf shrubs dominating large areas. This arguably virgin land could be settled without risking conflict with significant existing populations, offering parallels with New Zealand, probably the last major landmass to be settled *de novo* by humans. Not only were there no native people, but there were no grazing mammals either.

When the first Nordic settlers arrived in Iceland, they brought their pastoral ways of life and livestock with them. The pre-settlement ecological environment was sufficiently resilient to recover from the impacts of intermittent volcanic dust deposition and climatic change. Early settlers exploited the fragile subarctic-to-boreal ecosystems of this 'pristine island ecosystem', but the island's volcanic origin, erodible soils, cold climate and relative isolation made it highly sensitive to human impact. Numbers of people and livestock multiplied rapidly, and the introduction of the north-western European agricultural system of domestic mammals and crops rapidly transformed plant communities never previously subject to mammalian grazing pressure. Novel imported land-use practices including animal husbandry, extensive clear-cutting of natural woodlands and the introduction of numerous European insect and plant species wreaked severe and sustained impacts upon Iceland's fragile ecosystems. Subsequent centuries of agricultural use under Iceland's marginal natural conditions caused a shift to ecosystem degradation, large-scale soil erosion and desertification.

A large proportion of Iceland's vegetation cover has been lost during the past 1,150 years. Since the first settlement, most of the country's original trees and vegetation have been destroyed, and about half of the soils have eroded, resulting in barren deserts in many areas. Wood was so scarce that two eleventh-century Norwegian kings, Óláfr Haraldsson and Haraldr Sigurðarson, are said to have gifted timber (and a bell) for a church at Þingvellir. Human-driven processes of erosion impacted upland areas and subsequently encroached upon lower regions, and when market demand during the nineteenth century shifted from milk to meat, a substantial increase in sheep numbers grazing in the Highland boosted soil erosion and desertification there. Agricultural land use reduced the stability and resilience of the natural systems in the face of natural disturbances, such as volcanic eruptions and the cold spells of the Middle Ages. Here it is important to recognise and acknowledge that not all the deserts of Iceland were created by humanity's overexploitative behaviour, for many can be considered as natural deserts, at least to a degree, particularly those at higher altitude. But nevertheless the human failure to manage natural resources appropriately caused loss of resilience and large-scale landscape changes. Large areas of Iceland that were green at the time that Vikings landed are now lifeless brown deserts lacking any signs of human presence. *Landnám* had a powerful and lasting impact on the Icelandic environment, making land degradation a major factor shaping Icelandic ecosystems and one which has become a major contemporary environmental concern.

Much of Iceland's most severe soil erosion occurs in the deserts, but large tracts of the remaining vegetated areas are also subject to or threatened by erosion. The black-sand deserts are some of the most active aeolian areas, and the most serious erosion is found here. Sand encroachment is one of the most destructive forms of land degradation in the country, historically devastating large areas that were formerly fully vegetated ecosystems. There are two main causes:

- straightforward sand-blow from sandy areas and deserts onto vegetated expanses, which accounts for most of the encroachment; and
- the novel phenomenon of *sand tongues* or *fronts*.

When aeolian processes transport volumes of friable volcanic sand at ground level, it readily invades vegetated and cultivated areas in the form of sand fronts, tongue-shaped streams of sand actively swarming outwards from the deserts. These frankly weird natural manifestations invade, abrade and inundate vegetation, destroying ecosystems and fertile soils alike and leaving barren deserts in their wake. Vaguely reminiscent of the island's glaciers in their elongated linear and invasive form, sand fronts move as a continuous flux of sand and may extend up to 6 km in width and over 10 km in length. Capable of advancing rapidly, extreme instances have advanced up to 300 m in a single year – and even 300 m *over the course of a single storm*. A continuous flux of sand deeply erodes the soil surface, producing a wholly new surface which may be 1–2 m lower than the original. Fertile soils and vegetation are replaced by infertile sands, the land's productive capacity declines, and the species composition changes dramatically. Something as seemingly innocuous as sand has caused immense ecosystem damage change in Iceland, causing widespread desertification across large areas of south and north-east Iceland. Arnalds recounts how in Iceland a soil of 1–2 m depth on a fertile tract of land formerly covered with birch woodlands and harvested for fuel-wood into the 1600s was systematically eroded by wind and water erosion, leaving a barren gravel surface. The degradation of this area of common land was driven by land use: sheep grazing.

Large-scale and sustained soil erosion is intensified by a simple lack of water in many Icelandic deserts, despite the prevailing humid climate. The coarse-textured nature of the surface material, its unconsolidated and unstable nature, and the lack of significant organic content not only makes it susceptible to wind and water erosion but also leads to high water infiltration rates and high permeability, resulting in exceedingly limited water retention. The Icelandic deserts are frequently parched, and arid soil is much more readily eroded by the wind than moist soil. The lack of available water is one of the key factors constraining plant growth, alongside soil infertility, and the consequential lack of vegetative cover further exacerbates the deficiency of water. These arid desert surfaces may have as little as 1–2% plant cover, and that frequently only located in sheltered depression areas where moisture may be more available. Furthermore, the dark colour of the deserts translates into a low albedo, too, meaning the ground surface absorbs more heat than it reflects. Surface temperatures in this location close to the Arctic Circle may attain 35 °C or more on sunny days, an extraordinary figure which serves to accelerate evaporation, and frequent strong and dry winds desiccate the ground surface yet further.

Notions of the deserts being otherworldly are reinforced by NASA's selection in the mid-1960s of the barren Icelandic countryside for training the astronauts preparing for the first Moon landing, because of its resemblance to the surface of the Moon. Contrary to popular belief, however, Iceland wasn't selected primarily for its landscape but for its geology, providing field training for Apollo astronauts for their work in rock sampling, specifically on which geological samples were required. The basalt rocks and volcanic geology of Iceland's barren highlands were assessed to be the most Moon-like of any available training area. The various rock assemblages found in glacial outwash channels resembled those of the debris layer on the lunar surface, set in a landscape similar in appearance to the 'magnificent desolation' of the Moon, as Apollo 11 astronaut Buzz Aldrin was said to have remarked upon landing. Apollo 11, 12, 15 and 17 explorers were also introduced to the range of newly formed volcanic geology and associated rocks and structures found in Iceland. More recently, the prototype of the Mars 2020 Rover landing vehicle was tested in Iceland prior to its 2021 mission, collecting samples of rock. Iceland's deserts are truly otherworldly.

FLORA AND FAUNA

The Icelandic deserts support a depauperate range of flora and fauna. An inhospitable rock located in the middle of the North Atlantic Ocean, the island is set at a northerly latitude and, though mild given its location, the climate is nonetheless relatively harsh compared to elsewhere in Europe. Ecosystems are relatively simple, and vegetation characterised by low-growing species, with erosion leading to loss of vegetation cover being the chief threat to terrestrial flora and predominantly low temperatures, making for an environment of low biological productivity. Also, Iceland has for long been isolated from other landmasses, making it difficult for plants and animals to disperse to the island. Iceland's terrestrial biological diversity is thus low, lacking the relative complexity of other European terrestrial environments. There are few endemic species of fauna and flora, and notably few species of vascular plants, in contrast to the great diversity of marine fish species in the nutrient-rich and highly productive waters around Iceland.

The only native tree forming woodlands is birch, which at the time of settlement may have covered 25% or more of Iceland, though most is shrub-like in habit and less than

Reindeer in snow. (Jónína Guðrún Óskarsdóttir)

2 m in height. The single indigenous terrestrial mammal is the Arctic Fox, though Polar Bear occasionally visit Iceland, travelling on drift-ice in years when it reaches the country. While seabirds, waders and waterfowl represent one of the most important elements of Iceland's bird fauna, as few as 75 bird species nest regularly in Iceland. A distinctive feature of the fauna of Iceland is the complete absence of reptiles and amphibians, and the eighteenth-century account *The Natural History of Iceland* by Danish lawyer Niels Horrebow is lauded for its seventy-second chapter, 'Concerning Snakes' which, in its English (perhaps mis-)translation, comprises only one sentence: 'No snakes of any kind are to be met with throughout the whole island.'[3] This pointed observation founded a once-famous literary anecdote about 'the book with the chapter anyone can recite from memory' – but the passage is best interpreted as indicating simply that snakes had not been introduced. Curiously enough, snakes and indeed most non-native reptiles are illegal in Iceland, having been banned not only to preserve native ecosystems but also to avoid public health risks, for there have been instances of Icelanders contracting serious illnesses from reptile pets.

These are novel and fascinating desertic environments, but most of the Icelandic deserts are in the remote interior, and inaccessible other than by 4×4 vehicles or on foot or horseback. Few visitors seek to visit them: requests made to the tourist office to help arrange a field visit and a deserts tour drew a polite but – perhaps predictably – fruitless response. Popular natural environment tours head for glaciers, waterfalls, hot springs, volcano cones, ice caves and the 'Northern Lights' (aurora borealis), but the Icelandic deserts are not a regular tourist attraction. On the coast, the black-sand beaches of this North Atlantic desert island attract visitors drawn to their novelty, but many Icelandic coastlines are surprisingly dangerous, with freezing water, powerful ocean

---

3   Karlsson, G. (2000) *The History of Iceland*. Minneapolis: University of Minnesota Press.

ICELAND: COLD BLACK DESERTS

Heath Spotted-orchid – found among grasslands, heaths and birch stands of the lowlands and coastal mountains. Flowers range from white to pink in colour. The Icelandic name of this species is Brönugrös. (Holger Krisp, CC BY 3.0)

Rowan tree with berries. (Anne Burgess)

Arctic Fox on the hunt. (Erik F. Brandsborg, CC BY-SA 2.0)

The Atlantic Puffin breeds in Iceland – Iceland and Norway together account for about 80% of the European population. (Jakub Hałun, CC BY 4.0)

The Merlin is the commonest bird of prey in Iceland. It is the smallest falcon of Europe, though a fierce one, preying mainly on birds in flight. (Travis Williams)

A pair of Ptarmigan in the snow. A common breeding bird in Iceland, its favoured habitats are heathland and grassland. The species forms a significant part of the winter diet of the Arctic Fox. (Andrew Mckie)

currents flowing close to the shore, high and gusty winds making for formidable waves, and steeply shelving beaches with a powerful undertow. Numerous people have died, including at Reynisfjara, a well-known black-sand beach fringed with striking angular columnar basalt cliffs which attracts over 100,000 visitors each year despite being the most dangerous beach in Iceland. The tourist board advises against *even turning your back* to the ocean while on the beach there – let alone bathing or surfing – for just standing near the water can be dangerous, as huge 'sneaker' waves snatch the unwary and drag them out to sea, even on seemingly calm days.

Better perhaps to head for the human-made beach at the Nauthólsvík mini-resort on the coast at Reykjavík, where hot geothermal water is pumped into an artificial semicircular bay fringed with imported yellow sand, a seaside paradise making swimming in the North Atlantic Ocean available to all. The island's famed volcanism has other benefits, too: in the village of Friðheimar, hot water tapped via a borehole provides water at 95 °C, supplying sufficient heat to huge greenhouses to produce 370 tons of tomatoes annually – over 15% of Iceland's tomato consumption – and throughout the year, even in the coldest of the winter months. Almost all of the country's electricity is derived from renewable sources too, including geothermal, hydroelectric and wind energy, and most of Iceland's housing is heated by geothermal sources.

DESERT DUST

Sand is an enemy in Iceland, not only in terms of the invasive variety at ground level but also in terms of public health. The phenomenon of fine desert dust particles being blown around the country on such a scale inevitably gives rise to hazards for humans as well as species and ecosystems. The powdery nature and tiny size of volcanic sand means that the threshold wind velocity required to activate these materials is low. Dust storms represent a severe form of accelerated soil erosion, and the country experiences about 135 'dust days' each year, a frequency comparable to major desert areas of the world including Iran and the Gobi in Mongolia, and much higher than the USA. Dust storms are quite common in the vicinity of many of the major deserts along the southern coast. Anyone unfortunate enough to experience one will recognise its similarities to fog, as both light levels from the sun and visibility diminish drastically and often across thousands of square kilometres. Some dust hotspots in permanently windy southern Iceland are active all year-round, meaning that winter may bring combined snow–dust storms and maybe even black snow too.

Deserts contribute to air pollution around the world, and several studies have concluded that suspended desert dust can increase mortality hundreds of kilometres downwind. Iceland's capital city is located downwind from three active dust sources, and dust storms driven by strong winds bring substantial quantities of material in suspension: this 'Reykjavík haze' causes air quality to deteriorate, incongruous in a country accustomed to some of the lowest levels of air pollution in Europe. Desert-derived air pollution is in the form of high concentrations of ultrafine PM1 particles, the smallest-sized dust fraction with a grain size of a micrometre – one thousandth of a millimetre or less, about the length of a typical bacterium and thinner than the width of a single strand of spiderweb silk. Such submicron particles are capable of travelling much longer distances than larger particles.

Inhalation of desert dust emissions can have negative effects on human health. Exposure to heavy falls of fine ash can exacerbate asthma and bronchitis. The smallest and most mobile particles are of particular concern because of their potential for penetrating the lower respiratory tract and entering the bloodstream, where they can affect internal organs and be responsible for cardiovascular disorders. Mortality may be increased hundreds of kilometres downwind from dust sources. Some infectious diseases are associated with desert dust transmission too, including pneumonia, COVID-19 and pulmonary tuberculosis. Dust events are frequent in Iceland, where the volcanic material differs from that originating from most continental deserts, composed as it is of a higher proportion of submicron particles than crustal dust. Typical storms may transport several hundred thousand tons of dust, impacting areas of several thousand square kilometres. High PM1 concentrations in some areas of Iceland cause levels of air pollution comparable to those affecting heavily industrialised regions of Europe and Asia. Composed of about 80% volcanic glass, the material is angular in nature with sharp-tipped shards. Negative effects upon human health are worsened by the tubular morphology of the fine glass grains, which is not dissimilar to that of asbestos. Daily atmospheric concentrations of particulate matter in the Reykjavík air exceed health limits during some dust-day events, and are worse closer to the dust sources where concentrations may exceed health limits by multiples of 10–100 times.

The impacts of Icelandic desert sandstorms extend to other facets of life in ways mostly unknown elsewhere in Europe. The public is advised to stay indoors. Car drivers avoid travelling during intense storms, for there is little shelter to be had in the Icelandic countryside, and high-velocity winds may propel unwary drivers' vehicles off the road. Vehicles are consigned to the garage, or parked on the leeward side of the house. A regularly updated website advises drivers on a range of natural hazards which may be encountered on their journey, including blowing sand alongside more conventional ones like ice, snow, storms, blizzards and fog. During storms, abrasive wind-whipped sand and gravel can cause severe damage to road vehicles, a problem potentially so serious that Icelandic car hire companies recommend their rental clients take out optional additional insurance: Sand and Ash Protection provides cover for the car's paint, plastics and chrome being sandblasted, and auto glass becoming opaque and even broken by larger chunks of rock materials during storms, the repair costs for which might amount to several thousand pounds. Wind-wrenched doors are another potential hazard: drivers are advised against opening the doors on opposite sides of their car simultaneously during high winds.

HUMAN HISTORY

In common with many other desertic places, Iceland has been a place of exile, refuge and uncivilised lawlessness. At and after *landnám* during the Viking Age, the island served as a place of refuge for elements of the Nordic population. Nearly all the first Icelanders are thought to have emigrated there to escape the Norwegian monarchy at a time when, in the twelfth and thirteenth centuries, the first King of Norway, Haraldr Fairhair, was said to have unified the country into a single kingdom. Many Norwegians were driven into exile, including several minor rulers along with assorted outlaws and exiles who also fled the country. As a new frontier, Iceland was a natural place to resort to for those needing to find new homes and lives, and perhaps too a place of refuge for unconverted traditional believers in a Christianising society. Also in common with several other European deserts, the harsh desertic interior of the country served as a place of internal exile. There is a long Icelandic tradition of outlawry reaching back to the time of first settlement, and according to legend outlaws and bandits would flee to the Highland to escape the law. There were no prisons in the country until 1770 (a building which now houses the Prime Minister's office), meaning that sentences handed down to lawbreakers would comprise fines or, for more serious crimes, banishment. Sheep traditionally roamed free in the mountains over the summer, and stealing sheep was punishable by death. From the poem of the well-known Icelandic writer and Romantic-era poet Grímur Thomsen, 'Á Sprengisandi' (c.1875; now one of the most famous Icelandic folk songs), comes the verse:

> Outlaws in Ódáðahraun,
> Are maybe secretly herding [stealing] sheep

People who had been declared outlaws could be legally killed, and some notable Saga figures were put to death; indeed, it wasn't unknown for relatives of their assumed victims to commit the deed themselves. Icelandic farmers feared these men of the mountains with nothing to lose, bands of outlaws living by robbery and theft and murdering without compunction. The name Ódáðahraun, in fact, roughly translates as 'mischievous lava field', or 'bad deeds lava', and it is also known as the 'desert of the outlaws'.

To avoid such a squalid demise, outlaws like Eyvindur of the Mountains, the most famous of all, might spend years in the wilderness. In 1746, at the age of 32, Eyvindur was accused of theft and fathering an illegitimate child. Determining to take flight, he spent 20 years or more as an outlaw and fugitive in the wilderness with his wife Halla. It would have been a punishingly hard life living in the middle of a vast highland desert, killing sheep and living off the land, and he was said to have escaped from captivity on one or more occasions – though legend has it that Eyvindur only took the sheep of men who had wronged him while he was a free man. The couple are thought to have returned to civilisation in their eighties, and what are regarded as his bones were discovered in the ruins of a small farm and subsequently carbon-dated to the mid-eighteenth century. The visitor today can view the site of a cave he was said to have lived in, and at Herðubreiðarlindir he lived in a small hole in the ground covered with a horse carcass, a time he declared to have been the worst winter of his life. Outlaw tales have probably been exaggerated, instilling a perhaps disproportionate dread of what were usually probably no more than common sheep thieves, but expeditions would be mounted to capture them or destroy their hard-earned food stocks and supplies. Curiously enough, and perhaps instructively, perceptions changed in the nineteenth century as outlawry faded into history and, inspired by national Romanticism, Eyvindur and Halla went from zeroes to heroes, becoming the subjects of popular plays and even a Swedish silent film in 1918, *The Outlaw and His Wife*, ranking them – however incongruously – alongside other celebrated heroes of the Sagas as icons of Icelandic culture.

The litany of adversity faced by the inhabitants of Iceland has meant that many were forced to contemplate emigration. In the latter quarter of the nineteenth century, 20% of the population left for Canada and the United States. Not only was Iceland a largely desertic country but also one subject to frequent volcanic natural disasters and disease epidemics, blessed with relatively few valuable resources, offering few career and life opportunities, and stricken by poverty. Contemporary accounts from the eighteenth and nineteenth centuries tell of cheerless little hovels made of bare blocks of lava: single-storey houses roofed with furze (gorse) and twigs, and covered with turf. Timber was so scarce that whale ribs frequently served as rafters; floors were earthen; and the only seats were the bones of a whale or a horse's skull. John Ross Browne's description of houses at Þingvellir in 1867 was short and sharp: 'I can not conceive of more wretched abodes for human beings.'[4] Typical family houses were small, and inside the aroma of fish and sheepskins was so pungent that 'the marvel of it is that they don't come out [...] wagging their fins or bleating like sheep'. Nevertheless, Browne, an American travel writer and sometime US government official, observed in his book *The Land of Thor* that the priests studied the classical languages in these 'rude habitations', perfected themselves in the early literature of their country and became learned, devoting much of their lives to the pursuits of science.

The nineteenth-century plan for the complete abandonment of Iceland as a response to the poor quality of life there is a sobering one. In 1874 a 24-year-old Icelander journalist appeared before US President Ulysses Grant to outline a proposal to move the entire population to an island off the coast of Alaska. This radical notion was for a reformulated Icelandic nation, to be achieved by relocating 70,000 people to what was envisaged as an exclusive settlement in the New World closed to non-Icelanders. Such notions of abandoning Iceland completely were unrealistic, of course, and the starry-eyed plan came to nothing as it became apparent that there was no interest at the US government level, not least because of the likely cost of any such project to the US Treasury. But the episode does perhaps confirm historic domestic perceptions of the country as being a marginal one, both in geographical terms and also in the sense that Icelandic society may have verged on extinction during some of its most impoverished times. The notion of mass planned emigration as a response to the plight of living in one of Europe's most challenging desert regions is a particularly grim one.

Iceland's areas of wilderness have inspired numerous superstitious tales and legends. While there are precious few accounts of the Icelandic deserts in literature, the island's inhabitants have told folk tales of 'little people' since the twelfth century, and today 70% of the population is said to believe in the existence of a *huldufólk*, a 'hidden people' or 'hidden world', of elves, pixies and spirits. Thomsen's 'Á Sprengisandi' poem and folk song describes how a rider venturing into the desert may encounter not only outlaws but mythical creatures, spirits and an elfin queen too. The *huldufólk* possess magic powers, and are not necessarily by nature benign or well disposed towards humans. Crossing elves is considered highly inadvisable, and a road constructed not too far from the capital

---

4   Browne, J.R. (1867) *The Land of Thor*. New York: Harper & Brothers.

features a seemingly arbitrary bend in its course, because the roads department had been advised by local mystics to avoid bulldozing an area of volcanic rock where elves lived to avoid them exacting revenge upon humans for the disturbance. Many uninhabited places in Iceland are considered to be haunted by ghosts and witches, including Mýrdalssandur, 70,000 ha of desertic, barren black glacial sand outwash plain at the southern tip of the island, which was farmed from around the year 1000 but mostly abandoned by the fifteenth century in the face of eruptions and *jökulhlaups* cascading from the Katla volcano 35 km inland. Partly ice-covered and still active, Katla is being closely monitored, for the volcano is said to be statistically due for a new eruption.

Not far from Mýrdalssandur is the coastal desert of Sólheimasandur. Here lies the wreck of a US Navy C-117D military transport aircraft which crashed in 1973, an incident said to have been brought about when the pilot incorrectly formed the impression that the plane had run out of fuel. Happily, he had the fortitude to successfully crash-land the aircraft onto the matt-black barren Sólheimasandur plain with no casualties among the seven crew. The aircraft has remained relatively intact, and its unsalvaged empty fuselage has subsequently become a minor tourist attraction; but a sometimes ill-fated one, for most tourists reach the wreck on foot across the flat stony sands. It is easy to get lost in the area, however, and several have had to be rescued while attempting the two- to three-hour return journey during deteriorating weather. At least three have died in the Sólheimasandur desertic wilderness.

The former volcano Maelifell stands 200 m above its surroundings of Mælifellssandur in the Highland of Iceland near the glacier Mýrdalsjökull. (Tomáš Malík)

Solheimasandur aircraft wreck. (_Explanders, CC BY 2.0)

HALTING THE ADVANCE OF THE DESERTS

Out of necessity, Iceland has had to take action to restrain the spread of its unruly and invasive deserts. Much damage occurred during a cold period from 1860 to 1890, when high winds sweeping drift sand onto farms in central southern Iceland caused many to be abandoned. Centuries of agricultural use in Iceland under marginal natural conditions have caused severe and large-scale land degradation and desertification which threatened the existence of several communities. In response, in 1907 the country established perhaps the first state agency to conserve soils, before even the much better-known USA Soil Conservation Service (from 1933; now the Natural Resources Conservation Service). The land surface of the country has been mapped according to its erosion severity, with a National Soil Erosion Database classifying areas on a scale from 'No Erosion' through 'Considerable Erosion' to 'Very Severe Erosion'. Approaching half of the country's land surface is classified as subject to 'Considerable' to 'Very Severe' erosion, and overall it is reckoned that the scale of the issue in Iceland is comparable to the most severely degraded arid areas of the world. Today's Soil Conservation Service of Iceland collaborates with farmers, local government and NGOs on revegetation and ecosystem restoration projects – which include fencing, seeding grass species, and establishing and restoring natural woodlands. Over the first 75 years, the advance of the deserts was halted and even reversed, with some of the most severely affected areas having been reclaimed for agricultural use. However, large areas remain vulnerable to further erosion, and despite some successes the projects were often addressing symptoms rather than causes, a key one of which was unsustainable grazing management. Sheep production in Iceland is heavily subsidised, and at rates among the highest in the world, providing the majority of income to some sheep farms – not least perhaps to maintain rural employment in some locations. Links between agricultural policy and soil conservation issues are being addressed by coupling sheep subsidies to land-use practices. There has been a shift towards more participatory strategies, community involvement, and ecosystem management for multiple benefits, and long-term exclusion of grazing can have positive effects, but only limited change seems to have been achieved over recent decades.

Planting flowers was another plan to reduce topsoil degradation and loss, but this less conventional notion suffered under the law of unintended consequences. The lupin flower was deliberately introduced to Iceland on a large scale in 1945 to facilitate land reclamation for reforestation and the prevention of erosion along roadsides. A tall, perennial, pea-family herb of North American origin capable of growing to 1.2 m in height, Nootka Lupin is a pioneer species. In a country where it is difficult to grow anything, nitrogen-fixing Nootka soon became firmly established as a feature of Icelandic landscapes and gardens. This exotic plant proved so vigorous that it has spread largely unchecked and now covers extensive areas, having become widespread across south and north-east Iceland. Lupin has been declared an invasive alien species by the Ministry for the Environment, and nicknamed 'the Alaskan wolf' by others, a more caustic term punning on the similarity between the word lupine (wolf-like) and the flower species. With the ability to produce dense vegetation cover and improve soil fertility within a

Nootka Lupin in flower. (Anjali Kiggal, CC BY-SA 4.0)

Woolly Fringe-moss – in Iceland, *Racomitrium lanuginosum* is a pioneer species that colonises lava rock surfaces. (HermannSchachner, CC0, via Wikimedia Commons)

relatively short time and at low cost – as fertiliser is not needed – the lupin has achieved a degree of success in reversing land erosion, and its incongruous-for-Iceland flamboyant purple and white flower spikes have proved highly attractive to summer tourists. However, the species has invaded the south coast and some of the lower-altitude Highland areas and is regarded as a 'floral hazard', displacing native vegetation, forming extensive and dense monocultures which prevent other plants from growing – including the Bilberry, for instance – and more broadly resulting in a loss of plant species diversity as well as a change in community composition. Sensitive floral species including distinctive Icelandic mosses are outcompeted, and Iceland's only native bee – the Heath Bumblebee – is under threat, as are breeding birds and habitats including riverbeds. Such exponential expansion rates are often exhibited by invasive species, and while in 2021 Alaskan Lupin was estimated to cover only 0.5% of Iceland, as much as 13% of the country's land surface may be suitable for the plant.

This plant is difficult to manage once established in an area; eradication measures have not always proved successful due to their labour-intensive nature and high cost. Nature conservation legislation now bans planting Nootka Lupin in many areas, including at altitude in the Highland interior which is currently too cold and dry for most plants. However, projections suggest that in about 30 years under the current rate of climate change, the plant may colonise many of the Highland areas presently providing habitat to native species. Not only are there concerns that legislation may not check the plant's continuing spread, but opposition has been expressed by both tourists and Icelanders at efforts to mow down some of the sweeping expanses of lupins, which offer an undeniably spectacular sight at midsummer in a country of monochrome vistas.

DESERTIC ICELAND

Such a small, remote and even marginal European country, yet so many superlatives:

- the world's largest volcaniclastic deserts
- the largest area of terrestrial wilderness in Europe
- a largely desertic landmass
- the largest desert in Europe, with arid desert accounting for more than 40% of the island
- large tracts of land subject to or threatened by erosion and some of the most active and prolific aeolian sand sources on Earth
- one-fifth of Iceland is covered by sandy deserts, whose scale and extreme environmental nature are unparalleled in Europe.

Arid desert on a scale unparalleled in Europe. Geologically young landscapes which are intensely dry and continually on the move. In Iceland, the natural environment is ultimately sovereign: its immense desert wildernesses remain impregnable. In common with so many of the subcontinent's arid tracts, they remain little known, mostly economically redundant, sparsely or unpopulated, barely vegetated and geographically remote. Their physical appearance certainly chimes with the original meaning of the term desert: desolate and/or abandoned. Iceland's deserts have much in common with those on the subcontinent, though perhaps nowhere else in Europe do they have a similar volcanic genesis. These wildernesses reinforce this book's contention that the delineation of deserts centres upon substrate rather than solely – or even necessarily – climate. One dissimilarity is that, while most other European instances are desertic enclaves typically set in surroundings of relative fertility, the Icelandic deserts are expanses of waste set in a country that is essentially largely wilderness in character: badlands in a bad land. Another is their profoundly grey-black appearance, strikingly different to the honeyed, sunwashed deserts 3,500 km distant. It is more than a little ironic that a major threat to their continued existence may be… an imported purple flower.

So Europe's deserts range from the hot to the cold; and also the very cold. Iceland is as much a desert island as are the Canaries, though a sub-Arctic Circle instance rather than a subtropical one, reaffirming – if it was necessary – that deserts are indeed distributed the length and breadth of Europe. Different climates, and at 4,000 km apart set at opposite ends of the subcontinent, but in many ways broadly similar environments. Yet most surveys of the world's deserts fail to acknowledge Iceland, or indeed often any outside the subtropics.

Ill-advised human decision-making leading to land degradation and desertification has been a major factor shaping Icelandic ecosystems and environments. The resultant issues remain a major contemporary environmental concern, and it is instructive that, in many parts of twenty-first-century Iceland, rural farming settlements continue to be endangered by anthropogenic soil erosion which began in the tenth century. It is hard to escape the conclusion that Iceland offers a cautionary tale about what may happen in many other parts of the world unless humans manage natural resources appropriately.

Chapter Seven

UNITED KINGDOM: BRECKLAND – AN ARID DESERT IN VERDANT EAST ANGLIA

East Anglia is a large rural English region lying to the north-east of the London and South East regions, the wealthiest chunk of the country. The satellite image (via Google Earth) shows a region that is mostly green, for eastern England is the UK's breadbasket, comprising lowlands with some of the best soils and climate, together offering an excellent growing environment for cereals including wheat and barley and also for horticulture and pig and poultry farming. Almost no motorways disturb the bucolic serenity across the historic counties of Norfolk, Suffolk, Cambridgeshire and Essex, and there are few major centres of population in a region with relatively few inhabitants apart from Norwich, the traditional regional capital of East Anglia and a cathedral city with a modest population of 145,000. To the east of the city lies the Broads National Park, Norfolk's best-known holiday location comprising over 60 areas of broad, shallow lakes and navigable waterways along seven rivers stretching to the North Sea coast. This low-lying area was extensively quarried during medieval times for peat to provide fuel, and over subsequent centuries the abandoned pits filled with water. Now a major wildlife haven and one of the country's few lowland national parks, the UK's largest protected wetland is a popular summertime boating destination and also a haven for scarce water birds – including Bittern, Crane and Marsh Harrier – eels, insects and a rich wetland flora.

Not too far south-west of Norwich, and right at the centre of the region, the satellite image depicts a large dark blob. This is Breckland, an environment as arid as the Broads are watery, and a biogeographical region of eastern England today covering about 120,000 ha. Breckland is in its way the UK's most distinctive European desert, one of the least-known areas of East Anglia and with some of the poorest sandy soils in the country. The dark blob is Thetford Forest, the largest lowland plantation forest in England. Here is the familiar signature landscape of plantation forestry, with its serried ranks of non-native pines making for an eerily sombre atmosphere and a monotonous ecology. Yet barely 100 years ago – before the tree plantations – the landscape here across the Norfolk–Suffolk border was wide open, presenting sun-washed arid vistas and expanses of sand dunes… 70 km inland from the sea! Today only remnants of the pre-plantation-forestry landscape remain across the region, but Breckland's present-day patchwork of pines, broadleaf trees and relict lowland heathland is dramatically different from the surrounding landscapes and the tract remains a flourishing biodiversity hotspot. Once a remote badland region where drifting sand threatened settlements and society, its appearance was so unfamiliar that it bewildered travellers and geographers, and it has regularly drawn comparisons with the Eurasian steppe. In this remote and unknown area, humans faced existential challenges unknown elsewhere in the region; some of the inhabitants had a bad reputation; some of the wildlife posed a threat to humans; and, improbably, for centuries the humble rabbit was king in Breckland, playing a crucial role in its landscape, economy and society.

ARID BRITAIN

The first thing to understand about Breckland is that it is arid. That alone establishes it as exceptional, for soils over much of rainy UK are poorly drained or waterlogged. In many areas excessive soil moisture has been identified as the primary constraint on agricultural land use, and for centuries land drainage was a vital measure to increase agricultural yields. The UK has been identified as the most extensively underdrained region of Europe, though such field drainage comes at considerable cost for landowners and farmers. But the desertic tract that is Breckland requires little or no drainage, for the geologically similar sandy soils here retain neither water nor organic matter, and the relatively dry East Anglian climate yields only limited precipitation. Low levels of groundwater availability and a soil moisture deficit make for sparse vegetation and a restricted range of plant species. The prevailing winds further serve to dry out the light, drought-prone and shallow soils on this gently undulating plateau, thus increasing their susceptibility to wind erosion, especially in spring, though this all suits the locally adapted wildlife well. This is a near-waterless territory bordered by rivers flowing away from its margins, and meres – ponds that formed in shallow depressions dating from former cold times which were once a valuable source of water for livestock. Soils are highly permeable, with rainfall usually sinking without trace, and have low water retention capacity, although water may be readily located below ground in places. The lack of available water limited the development of settlements on the Breckland plateau, leaving a pattern of villages concentrated on the sides of the shallow valleys, and small hamlets and farmsteads scattered over the drier central uplands. As water was so scarce, it was carefully allocated to the various parishes, which agreed between themselves who was entitled to what water. No fewer than nine parish boundaries converged at Rymer Mere, each with access to the water's edge indicating its importance to peasant farming. No wonder that Hoskins termed Rymer 'an English oasis in a dry land, like all the other meres'.[1]

The large arid territory that is Breckland centred on the ancient towns of Thetford and Brandon is identified by a unique combination of physical attributes, wildlife, land use and culture. Set on a low rolling plateau rising to about 60 m, the surface geology is of glacial outwash sands and gravels overlain with the most extensive deposits of windblown

---

1   Hoskins W.G. (1978) *One Man's England*. London: BBC.

coversand in southern England, to depths of 2.5 m or more. Breckland is interesting not least for perhaps representing the most significant western extremity of the arid coversand belt stretching across northern Europe. Also here in places is boulder clay deposited by retreating glaciers, and underlying everything is chalk, free-draining and often near the surface. Located amidst the East Anglian region, which receives only about half the UK average rainfall, Breckland experiences a climate relatively less oceanic-maritime and more continental than the rest of the UK, with slightly more pronounced extremes of temperature. The area is slightly drier and with long sunshine hours; relatively high summer temperatures quickly give way to colder ones in winter; and air frosts may occur in all months of the year. Low precipitation, coupled with free-draining sandy soils and losses through transpiration and evaporation contribute to a significantly negative soil moisture budget. Breckland's physical attributes have led to the development of dry heath and grassland communities here and, in combination with land use and culture, have resulted in wildlife communities unique within the British Isles.

Flint once dominated extensive areas of ground in some locations. Occurring in beds in the underlying chalk, scattered flints are ubiquitous across the region. A naturally occurring sedimentary rock composed primarily of quartz, flint is harder than the chalk in which it lies and as it does not dissolve in water it remains in the form of irregular nodules once the chalk has weathered away. As a result, flint accumulated in some areas to the extent that the ground was dominated by nodules, producing extraordinary bare jagged pavements of impoverished and parched soil which were called 'stony Brecks'. Ringed Plover, a coastal bird of shingle beaches, was once abundant on these flint-covered expanses. At the beginning of the twentieth century an estimated 400 pairs were breeding on Breckland, remarkable today and serving as an indicator of the former extent of the most intensively grazed and bare shingle-like conditions in Breckland, though the bird's presence here has declined due to habitat loss. Flint was quarried extensively in the later Neolithic period, most famously at Grimes Graves, a prehistoric mining area in the centre of the region. This is considered to have been one of Europe's first industrial centres, with several hundred mining shafts dug as deep as 13 m into the chalk to exploit seams of flint with antler picks. The rock was highly valued in the Neolithic for its ability to be easily fractured, worked – or 'knapped' – for making stone tools and weapons with razor-sharp edges. In modern times, flint from the nearby Brandon area again became a valuable commodity as an essential component of the ignition mechanism for flintlock guns, creating a shower of sparks which ignited the priming powder. Flint knappers in Breckland supplied gunflints for the British army in the Napoleonic Wars, and as late as 1950 some 2,000 gunflints were made each day in Brandon for export mainly to Africa. Flint's usefulness is also reflected in its historic use as a building material in conjunction with brick or chalk throughout this region with no local building stone.

THE 'BRECKS'

When Neolithic agriculturalists came to the UK more than 5,000 years ago, Breckland was one of the first regions they colonised. Some areas developed into extensive open heathland. Subsequently the region's land-use history has been a dynamic and changing one, historically characterised by large areas of extensively grazed dry vegetation, including grass- or heather-dominated heath and areas of low intensity agriculture. Breckland was extensively cultivated in the Middle Ages when marginal land brought into production, especially when the climate was warmer between 1100 and 1250. In medieval times, the most productive lands were the 'in-fields' closest to each village while further away were temporary fields called 'brecks' or brakes, tracts of sandy heathland and gorse which had been 'broken' off the heath for short-term arable cultivation. These brecks, typically subject to common rights, were farmed on a rotational basis as a means of gaining some production from these unpropitious lands. Each was ploughed for crops a few times, perhaps once over ten years, although the poverty of the soil meant that a successful crop was by no means certain. This shifting cultivation mode of agriculture was typically adopted when crop prices were high, and once the soil's limited nutrients had been exhausted the breck was subsequently abandoned to lie fallow and revert to heathland while a fresh expanse was ploughed. Thetford historian and naturalist William George Clarke coined the regional name Breckland based on the term 'breck' as late as 1894. Life here was hard for farmers historically, with some of the least inherently fertile soils in the country and a lack of rain. But while throughout history the Breckland heaths were regularly referred to by some as wastelands, in practice they were not 'wasted', as the grazing they provided was vital to the many small-scale farmers with little or no land of their own. There was a measure of freedom to be had here too, for a life of subsistence on the commons and wastes was deemed more desirable by some than the prospect of a life of wage-labouring at the beck and call of farmers. However, over-cultivation and over-grazing on the fragile Breckland soils led inevitably to their degradation and consequent wind erosion from the end of the thirteenth century onwards.

Sheep were the principal livestock. In the twelfth century, most of the large flocks of sheep numbering up to two to three thousand were owned by the religious houses which owned much of Breckland. Sheep-corn farming was an ancient practice in Breckland. This was a mixed farming method for boosting the fertility of arable land, with sheep grazing the open lands by day before being folded onto arable stubble or ploughed fallow land at night, where their dung fertilised the light sandy soils increasing crop yields. After the Dissolution of the Monasteries in the 1530s, the monastic lands were bought by local families. Most small farmers were unable to compete and moved away, with the result that, by the end of the seventeenth century, Breckland was less populated than at any time since the Domesday Survey. Large estates were established in the eighteenth century as wealthy farmers bought out their impoverished tenants. Extensive areas of heath were 'reclaimed' for cropping and landscaped parks. Many open areas of Breckland were enclosed by Act of Parliament in the late eighteenth and early nineteenth centuries. A few small heathland commons survived, but elsewhere the rights of commoners to graze their animals ceased and numerous roads and footpaths were closed. What had been a frugal community cutting heather for thatching and turf for the fire, and raising thin crops of wheat in their cottage gardens, was cut off from its resources. Eighteenth and nineteenth century enclosures resulted in large, regular fields often clearly defined by Scots pine and beech shelterbelts or hawthorn hedges indicative of large estate enclosure, and new farms and roads were built on former heaths. By the 1850s, the importation of cheap American grain made cereal growing unprofitable, and the Breckland was one of the first agricultural areas to become depressed, with farming ceasing in many locations and some land being let for shooting. Since the mid-twentieth century, much of the region has been transformed as land was brought into permanent agricultural use.

BRECKLAND LANDSCAPES

The remaining open landscapes of Breckland may look dry, empty and barren – and certainly unspectacular in any conventional sense – but closer investigation reveals an unusual environment with a huge diversity of life. This large, roughly triangular region is a biodiversity hotspot, supporting an extraordinary assemblage of rare and threatened species, some of which are found nowhere else on Earth. Much is classified as dry grassland; but also, remarkably, steppe, one of the world's great biomes yet almost completely absent from the UK. The steppe environment is one of vast dry lowland-to-submontane plains prevailing in temperate climates. Exhibiting little diversity in vegetation, steppe is dominated by plants adapted to hot summers with long periods of drought and cold winters, characterised by extensive grasslands with a few shrubs and too dry to support closed forests except near rivers and lakes. Roughly similar to the Great Plains in the USA, the Eurasian steppe is one of the largest continuous terrestrial biomes and the largest temperate grassland in the world, extending almost one-fifth of the way around the planet from its westernmost sizeable extremity in Hungary to China. More comparable to the Eurasian steppe than almost anywhere else in the UK, Breckland is the country's representative of this major world biome, although being located a long way

Great Bustard – a spectacular bird whose breeding range historically extended across the Eurasian steppe. It is now suffering rapid population reductions owing to the loss, degradation and fragmentation of its habitat, illegal killing, and general disturbance. (Ana Mendes do Carmo)

Wangford Warren rabbit burrows.

west of Hungary it constitutes an ecological phenomenon termed extrazonal or exclave. The rapid onset of a period of warm, dry conditions after the last Ice Age allowed raw chalk and sandy soils to be colonised by steppe and also some Mediterranean species. These were widespread in the UK in the warm and dry period after the last Ice Age, but were generally lost elsewhere as dense woodland developed while surviving in Breckland where forest cover may have been more open in character. Today rich in rare species, the remaining steppic areas are set in mosaics with heath and acid grassland, reflecting local differences in substrate. Once widespread in Europe, steppe is now a scarce habitat on the subcontinent – mostly due to agricultural intensification – and among the most threatened and least protected habitats globally. The magisterial Great Bustard, a bird of the open steppe, long favoured the treeless plains of the Brecks and clung on here into the middle of the nineteenth century when the last indigenous British examples were observed, one of the last wild breeding individuals having been shot in Suffolk in 1832. Extirpated through a familiar tale of habitat fragmentation and loss, tree plantation, agricultural modernisation and being shot for the pot, a population of Great Bustard was re-established on Salisbury Plain in Wiltshire in 2004.

The lowland heathland, grassland, scrub, lakes (Breckland meres) and forest environments that make up the Brecks provide habitat to nearly 13,000 species. This includes 26% of all UK conservation priority species, amounting to more than 2,000, over an area

Small Copper butterfly. (Stephan Sprinz, CC BY 4.0, via Wikimedia Commons)

Grayling butterfly. (Charles J. Sharp, CC BY-SA 4.0, via Wikimedia Commons)

Green Hairstreak butterfly. (Stephan Sprinz, CC BY 4.0, via Wikimedia Commons)

Sand Sedge is more usually found on coastal sand dunes. (Rutger Barendse)

of only 0.4% of the country. More than 160 species are restricted to or have major strongholds in Breckland, including 21 that are entirely restricted to the Breckland region. On hot summer days, the heaths teem with beetles, ants and solitary bees. Characteristic butterflies include Small Copper, Grayling and Green Hairstreak on the acid heaths, and Brown Argus and Dingy Skipper on the chalk. Heathland is a vitally important habitat for invertebrates, supporting many rare British species that are at the edge of their European range, and acting as home for 40% of British spider species. On the sands of Wangford Warren may be found Grey Hair-grass, a tough and nationally rare sand-dwelling species more usually found on coastal dunes, growing in dense tussocks here at one of only two English inland locations. About 25 species otherwise restricted to the coast are only found inland in Breckland, including beetles and bees. Many of these plants and animals are Breckland specialists, requiring dry open environments, a closely grazed sward, and disturbed ground. The Breckland region is protected with multiple legal conservation designations reflecting its varied biodiversity and landscape interest, including as Special Protection Area (SPA; for bird conservation) and Special Area of Conservation (SAC; habitat and fauna and flora conservation), high-level conservation designations denoting internationally important sites ranking among the best in the world and recognised as special places under international conventions. There are numerous Environmental Stewardship (more recently termed Countryside Stewardship) scheme agreements in place, which aim to support land management for biodiversity and landscape goals.

THE RABBIT IN BRECKLAND

Unlikely as it may seem, the history of the Breckland deserts is closely tied to that of the rabbit. By the late nineteenth century, the rabbit was king in Breckland, and the herbivore has had an incalculable impact throughout the British Isles and the numerous other countries around the world in which it has been introduced. Sheep fared tolerably on the arid expanses here, but the rabbit has been spectacularly successful economically and also in terms of its impacts upon biodiversity and environment. The parched dry desert sands of Breckland proved a particularly favourable environment for the rabbit, this blameless serene grazing mammal which still plays a pivotal role in the region's ecology and landscape – though one regarded alternately as both villain and saviour.

The European Rabbit is thought to have originated from the Iberian Peninsula and north-west Africa – so from a dry Mediterranean climate with relatively low rainfall, travelling all the way to the centre of East Anglia. First imported by the Romans, the rabbit was subsequently brought by the Normans in the eleventh century. A high-status commodity, rabbit was then a luxury item reserved for the upper classes. The adult was called a coney while young animals under one year of age were known as rabbits, and most prized for their meat. In the thirteenth century, one rabbit was worth more than a workman's daily wage. Rabbits were on the menu for Henry III at Christmas in 1253, Henry VI had a nightshirt lined with black rabbit fur, and the Breckland monks of Thetford Priory gave Katherine of Aragon a gift of rabbit fur trimmings for her gowns when she visited in 1513.

Initially this thermophilic species found the damp English climate inhospitable. The rabbit can tolerate a range of conditions, but much prefers light, well-drained soil, and thrives in a closely managed rearing environment: the warren. This was a specially created enclosure offering a safe environment and protection from predators and bringing something akin to industrial production to the countryside. Ownership of a warren was confined to those with manorial rights. The earliest instances were probably unenclosed open warrens, especially on the commons, but most were artificial creations, open and treeless environments surrounded by banks or walls. Inside, banks of soil were piled up, providing the sunny south-facing slopes they so appreciate and encouraging them to burrow. For five centuries the majority of England's rabbits lived within warrens, where they flourished: born, nurtured and protected before being culled by the warrener. Once ubiquitous across England, the warren was a feature of the landscape significant enough to be depicted on early maps, and today the place name 'warren' occurs at 1,000 or more locations in the United Kingdom. From the late twelfth to the early twentieth century, Breckland was renowned for having perhaps the densest concentration of warrens, with in the eighteenth century more than 20 in this remote corner of England.

For nearly nine centuries the rabbit was intensively farmed on the infertile Breckland soils. Here was a cash crop able to put the worst tracts of land to economic use in a region that had rarely enjoyed much prosperity. They were not merely a cuisine item; demand for their fur grew over time, including for the felt hat trade. The rabbit was hunted at times too, though the creature was perhaps not as highly prized as quarry as other 'venison'. They breed prolifically, with does – females – capable of giving birth to six to eight young on five or six occasions over the breeding season from February to September. Output increased over the centuries, with medieval warrens culling up to 3,000 each per year, and by 1860 around 36,000 were slaughtered in Thetford alone. Manorial lords with their flock masters and rabbit warreners became the wealthiest and most powerful members of society.

Rabbits are voracious feeders, and are able to survive on poor-quality vegetation. Their diet mostly comprises grasses but they are able to eat anything green, including spiky gorse. Drinking very little, they have no need for supplies of running water. The intense grazing regime both within and without warrens gave this area of fragile soils a distinctive and frequently desolate appearance. Burrows riddled and undermined slopes and banks, making walking and horse riding hazardous and leading the fifth Earl of Albemarle to write of nineteenth-century Breckland: 'The whole country is a mere rabbit warren, and still goes by the name of the holely [Holy] land.'[2] Warren erosion was adjudged so severe that in 1550 the people of Freckenham to the south-west of Breckland determined to slaughter all the rabbits on the common land of the village. Rabbit grazing has been implicated as a direct cause of sandstorms becoming regular events, as the uppermost layers of sand were exposed to the force of the wind leading to the formation of mobile dunes. In 1809, the height of the soil surface of Low Warren at Eriswell in Suffolk was lowered almost one foot (0.3 m) by wind action.

---

2   Sheail, J. (1971) *Rabbits and their History*. Exeter: David & Charles.

'View on Thetford Warren, illustrating the open and bleak landscape there' from the 1866 book *The Birds of Norfolk, with Remarks on their Habits, Migration, and Local Distribution* by Henry Stevenson and Thomas Southwell. (Natural History Library for Norfolk Museums Service in Norwich)

Rabbit escapees were first recorded in the wild in the fourteenth century. Frequently regarded as a rural pest outside captivity, feral rabbits sparked continuous complaints from neighbouring farmers over the following centuries. An inquisition held at Methwold in 1522 declared: 'Much of the Corne of the said londe is distroyed yerely with Conyes (conies) which be so greatly encreased.'[3] In Thetford, Westwick Heath was described as barren in 1601 because of rabbits and sheep. But not until the eighteenth century did rabbits colonise the wider countryside on a large scale. The Duc de la Rochefoucauld, a French aristocrat and sometime travel writer, wrote in 1784:

> A large portion of this arid country is full of rabbits, of which the numbers astonished me. We saw whole troops of them in broad daylight; they were not alarmed by noise and we could almost touch some of them with our whips.[4]

Over the centuries, the rabbit's status descended the social scale as it became more available, eventually becoming a staple of the diet of the poor. Theft by poachers was a constant threat. Usually situated in remote areas, larger warrens had a specially

---

3   Sheail, J. and Bailey, M. (1996) The history of the rabbit in Breckland. In P. Ratcliffe and J. Claridge (eds) *Thetford Forest Park: The Ecology of a Pine Forest*, pp. 16–20. Edinburgh: Technical Paper 13, Forestry Commission.
4   Rochefoucauld, Francois Duc De La (1933) *A Frenchman in England 1784*. Cambridge: Cambridge University Press, p. 212.

constructed lodge which served as home and workplace for the warrener and was fortified against gangs of marauding poachers. Some are still visible in the present-day landscape.

The rabbit has played a rarely recognised yet pivotal role in maintaining the landscapes and biodiversity of the East Anglian deserts. The rabbit is a keystone grazing species and saviour of biodiversity by virtue of its favourable impact on the ecosystem structure and biodiversity of this habitat. The influence of the creature upon the Breckland landscape has been nothing short of transformative, giving rise to one of the most extensive areas of lowland heathland and lowland acid and calcareous grassland remaining in the United Kingdom today. The rabbit's everyday behaviour centres around small-scale ground disturbance through grazing, burrowing, scraping and stripping turf. In these ways the creature engineers open landscapes, maintaining short grass swards, annual herbs, lichen- and moss-dominated areas, and expanses of bare ground, producing a varied mosaic or microtopography. Encouraging rabbits is essential to keep the habitat from succeeding to a closed grassland. Breckland has been subject to an intensive and centuries-long regime of nutrient stripping, notably including through the harvesting of rabbit carcasses which represented a substantial depletion of nutrients on warren sites. In the late eighteenth century up to 40 rabbits per hectare were harvested annually at one Breckland warren, and in the nineteenth century long-established warrens yielded poorer rabbit harvests than did the open lands. Other nutrient-stripping processes included livestock grazing, cropping and peat and gorse harvesting for fuel. Heather was used by the poor as thatch for house roofs, birch gathered for brooms, baskets and furniture, and bracken cut for bedding for humans and litter for livestock. All served to deplete nutrients – minerals – thus maintaining this English desert.

Diversity flourishes as many declining rarities including Breckland Thyme and Spring Speedwell require open, disturbed soil in which to germinate and become established – as do violets, Ground-ivy and Bugle, from which early-season butterflies derive nectar. Rabbits' grazing and digging activities also benefit the mosses, lichens and other plants that many characteristic heathland species – notably including such invertebrates as beetles, bees and butterflies – need to thrive. The rabbit's activities

Gorse – or Furze – is common in these habitats, producing masses of bright yellow flowers with a tantalising scent of coconut.

Breckland Thyme is a dwarf, aromatic, evergreen shrub. It is scarce but a characteristic component of the Breckland flora. (AnRo0002, CC0 1.0)

Eurasian Stone-curlew. (Frank Vassen, CC BY 2.0)

benefit the Stone-curlew, once common on Breckland, a curious-looking large bird with long yellow legs and big yellow eyes for nocturnal foraging. The UK's migratory Stone-curlew population declined for much of the twentieth century, but has partly recovered in response to an intensive conservation programme. Arid Breckland provides home for 65% of the UK's population of this steppe-dwelling bird, here close to the most northerly point of its range. Open patches of sandy soil created by rabbit activity warm up rapidly in the spring sunshine, providing ideal locations for the Stone-curlew both for basking and to serve as nest sites. The rabbit also fulfils a sacrificial role ecologically, in that it serves to distract predators such as foxes and stoats from feeding upon Stone-curlew birds and their eggs and chicks. As short swards of rabbit-grazed grassland have diminished, these ground-nesting birds are now to be found in spring-sown crop fields as frequently as upon sparsely vegetated ground, but the species is susceptible to disturbance and as their nests are difficult to spot some are unintentionally destroyed by farmers. They also provide prey for many characteristic Breckland birds including the Buzzard, whose usual meal is rabbit meat.

Warrening continued into the late nineteenth century, though fashions in cuisine and clothing had been changing for decades. Cheaper rabbit products were being imported from Europe, and profits fell locally as a result. Many warrens fell into disuse, and fur factories closed. Yet as late as the 1920s, 30 warreners were employed on a single Breckland estate, Elveden, where nearly half the area was warren and the annual bag amounted to 120,000 rabbits. Photographs of warreners from around 1925 show sturdy-looking countrymen with lurcher dogs at their sides and ferret boxes at their feet. The number in the wild increased prodigiously in the nineteenth century when the creature became a significant pest throughout the country. In the 1920s, agricultural land fell out of cultivation, and by the following decade wild rabbits had been identified as the UK's most serious vertebrate pest of cereal crops and grassland, at a time when dunes were forming on the unstable sandy soil of some Breckland tracts. Farmers fought a continuous battle against the creature, and on some fields growing winter-sown corn proved impossible. Breckland was overrun, with one estate erecting over 50 km of rabbit-proof netting from 1927 to 1930 before cropping could commence. The Prevention of Damage by Rabbits Act was passed in 1939, and by the 1950s the government estimated 60 million rabbits were causing £50 million worth of damage. However, in the 1950s the deliberately introduced viral disease myxomatosis killed over 99% of the wild population. The naturalist and author George Jessup recalled in 1954 walking 'over a heath where it was difficult to step between the dead and dying rabbits, and never in my life have I experienced any similar feeling of nausea'.[5] Breckland and similar areas became overgrown as a result, though rabbit populations recovered to a high point in the 1970s.

Today rabbit farming for meat continues though on a small scale, the UK importing much from China, Hungary and Poland. Managed warrens no longer exist in the UK, and many or most rabbits are farmed indoors in mesh cages. The advent of industrial fertilisers, agricultural machinery and irrigation systems eclipsed the rabbit's role as a wealth

---

5   Jessup, G. (1992) *Breckland Ramblings.* Wymondham: Geo. R. Reeve.

generator as Breckland soils could be farmed for potatoes, carrots, onions and parsnips. The viability and indeed meaning of a whole way of rural life quietly faded away. Today a few remaining warren banks and lodge sites are protected as ancient monuments, survivals of an industry which has vanished from the landscape. The national wild rabbit population has been in decline continuously since the 1990s, attributed to a combination of disease – including since the early 1990s the highly contagious and fatal rabbit haemorrhagic disease virus – persecution, predation and habitat degradation and loss. Myxomatosis still kills rabbits, too. As the rabbit has declined, heaths have become overgrown by coarse grasses and shrubs capable of outcompeting Breckland's less-competitive rare plants and insects, reducing overall biodiversity. Given their former dominance, it is interesting to speculate how the English countryside might look had the rabbit never been introduced – perhaps Breckland would have disappeared under woodland centuries ago.

SAND DRIFTS AND FLOODS

For centuries, Breckland comprised vast expanses of open sand and heathland where drifting sand threatened settlements and society. There are thought to have been five main Breckland drift-sand phases going back to about 5200 BCE, and those since 1100 CE have been particularly intense. It has been proposed that sand drift episodes over the last 2,000 years may have been linked less to climate fluctuations and more to changes in land management. The devastating Breckland sand floods may perhaps have been caused or triggered by the ill-advised actions of Breckland's inhabitants. The implication is that contemporary societies there were insufficiently aware of the threat posed by landscape instability, or that they failed to implement suitable measures to address the sand peril.

A combination of environmental factors and the impacts of human activities provided perhaps the ideal conditions in Breckland favouring extensive wind erosion and sand drifting over the centuries. Friable aeolian superficial sediments, a relatively dry climate, low relief and an open landscape combined with deforestation and the activities of the rabbit resulted in drifting sand and dune mobilisation. The region was historically disparaged by travelling writers and geographers for its remarkably unBritish desert-like appearance and a degree of agricultural inadequacy perhaps unparalleled in lowland Britain. John Evelyn, the English intellectual, writer and diarist, visited Euston – today a small green street village in south-east Breckland – in 1677, and wrote:

> The travelling sands, about 10 miles wide of Euston, that have so damaged the country, rolling from place to place, and, like the Sands in the Deserts of Lybia, quite overwhelmed some gentleman's whole estates.[6]

William Gilpin, English writer, clergyman and schoolmaster, is best known for his works on the picturesque, which he defined as 'a term expressive of that peculiar kind of

---

6   Cook, O. (1980) *Breckland*. 2nd edn. London: Robert Hale.

beauty, which is agreeable in a picture'.[7] However, Gilpin found nothing picturesque to report in Breckland. His vivid account of his experience travelling over the rough tracks crossing the sandy wilds during a 1769 tour of the eastern counties painted Breckland as a 'vast tract of land' and a 'disagreeable' one too. The landscape at the centre of the region was 'wild, open, and dreary'; not a tree was to be seen on the sands; and the line of the horizon was scarcely broken with a single bush. Sheep and cattle grazed on the greener margins of this landscape, but further into the desertic wilderness the limited available herbage diminished, and:

> In a few miles the country became an absolute desert. Nothing was to be seen on either side, but sand, and scattered gravel, without the least vegetation; a mere African desert [...] In some places this sandy waste occupied the whole scope of the eye: in other places at a distance we could see a skirting of green with a few straggling bushes, which being surrounded by sand, appeared like a stretch of low land shooting into the sea. The whole country indeed had the appearance of a beaten sea-coast; but without the beauties which adorn that species of landscape. In many places we saw the sand even driven into ridges; and the road totally covered; which indeed was everywhere so deep, and heavy, that four horses, which we were obliged to take, could scarce in [sic] the slowest pace, drag us through it. It was a little surprising to find such a piece of *absolute desert* almost in the heart of England. To us it was a novel idea. We had not even heard of it.[8]

Travelling further across this East Anglian desert, Gilpin related that the country was dry, and meagre. No animals were to be seen except a few rabbits, which are the only inhabitants it could support.

The existence here until recent times of mobile sand dunes is a geomorphological phenomenon more normally associated with the world's major subtropical deserts rather than the damp western European islands. Arid Breckland has a long history of drifting sand, and perhaps the best-known instance occurred at the village of Santon Downham in the centre of Breckland.

Thomas Wright, an inhabitant of Santon Downham, recounted in great detail his experience of a sand flood period there between the years 1665 and 1670, when the friable local sandy soils swept in. Downham Hall, Wright's large seven-bay brick house, was one of several to be almost engulfed in sand. Many other village houses were buried. The village's pastures and meadows, which would have been carefully husbanded for centuries after a medieval period of abandonment and were described as of good quality and extensive, were 'overrun and destroyed.'[9] What Wright described as the 'prodigious

---

7   Gilpin, W. (1809) *Observations on Several Parts of the Counties of Cambridge, Norfolk, Suffolk, and Essex*. London: Cadell and Davies.
8   Ibid.
9   Wright, T. (1668) A curious and exact relation of a sand-floud, which hath lately overwhelmed a great tract of land in the county Suffolk; together with an account of the check in part

sands' originated from a warren at Lakenheath, a tract of open country about 13 km to the south-west. Here the thinly vegetated soil surface of several 'great sand hills' was 'broken' by strong winds from the south-west, enabling sand transportation to commence. About four hectares of adjacent land was soon covered in this 'first eruption.' The mass of mobilised sand increased and continued drifting for 6 km, swamping more than 400 ha of land. In its path lay a single farmhouse, the owner of which first endeavoured to secure his home by building barricades against the sand, but to little effect: the 'winged enemy was not to be opposed', so he changed tactics. Instead of trying to prevent the invasion, he removed all bulwarks, fences and anything else that might obstruct the movement of 'this unwelcome guest', offering free passage. Within four or five years he was nearly rid of the unruly sand, so successfully that Wright was able to record that 'there is scarce any footſteps left of this miſchievous Enemy.' But, continuing its dusty progress, the 'sand floud' (flood) travelled downhill and uphill across a valley, advancing about a mile uphill over a period of two short months and inundating 'about 200 acres of very good corn land' until it reached the outskirts of Santon Downham, where its progress stalled for a time 'without doing any considerable mischief' to the village.

Then the sand commenced its invasion of this isolated hamlet. The Norfolk desert had, in Wright's words, turned into a 'conquering enemy'.[10] Houses were buried and destroyed; some roads were blocked to a depth of 3 m; pastures and meadows were swamped and rendered worthless. The modest Little Ouse river which flowed around the village filled with sand to such an extent that boats transporting goods could only carry a fraction of their former cargoes. Attempts by the hapless inhabitants to defend the village cost 'a greater expense than they are worth.' Wright himself was forced to make strenuous efforts to save his own house, for before too long the sand had blocked all approaches to the property, leaving no passageway except over two walls each about 3 m in height. His yard and a small grove of trees in front of the building were buried. Sand accumulated almost up to the roof eaves of his outhouses, and his garden wall collapsed under its weight. Over a period of four or five years, Wright countered the sand invasion as best he could. He continually raised ramparts of sand upon which he planted hedges of gorse, and as quickly as each was overwhelmed he raised a fresh one on top to a height of nearly 18 m, at some considerable cost. The bulwarks succeeded in confining the sand within an area of about 3.5 ha, into which he enterprisingly mixed hundreds of loads of 'muck and good earth' to restore the site's fertility over the course of a year. His neighbours helped to cut a passage to his house through the main body of sand, carting away 1,500 loads in a single month in an example of how human ingenuity and effort may triumph over the challenges presented by natural processes.

Wright's spirited defensive attempts seem to have been effective, showing that defeat by the invasive sands was not inevitable, even in a village once known by the perhaps

given to it; communicated in an obliging letter to the Publisher, by that Worthy Gentleman Thomas Wright Esquire, living upon the place, and a sufferer by that Deluge. *Philosophical Transactions of the Royal Society of London* 3(37): 722–725. London: The Royal Society. https://doi.org/10.1098/rstl.1668.0025

10 Ibid.

humorous nickname Sandy Downham. The distinguished English landscape historian W.G. Hoskins – who readily termed Breckland a desert – noted that remarkably little has been written about these places, once so common throughout England and Wales, or their peoples. Most such sand-flood incidents would have gone unrecorded: Wright's is a scarce contemporary account by someone literate describing the intensity, scale and pervasiveness of the historical British deserts and the threat they posed. Illiteracy is of course at the root of this lack of written sources, but also people not feeling they had a voice, or any sort of story to tell, and certainly not one about these rural wastes. Perhaps, too, the voices most absent are those of travelling peoples of all kinds, despite their centuries-long presence. Now surrounded by plantation conifers, there is little trace of the extensive sandy lands that once dominated the village's surroundings, nor of the historical tumultuous floods of sand. However, St Mary's Church bears a conspicuous single red brick set at a height of about 3 m amidst its grey-flint wall which, according to local legend, marks the height to which invasive sand accumulated during the seventeenth-century sand flood. The Forestry Commission built a new model rural village at Santon Downham to house its growing workforce, which by 1951 housed 287 people.

Breckland has a long history of human settlement, but also of village desertion, some instances of which may be attributed to drifting sand. Thought to have been one of the first places in England cultivated by prehistoric people, the region's landscape today is scattered with ancient circular features. Most of these are Bronze Age burial mounds which owe their preservation in part to the poverty of the Breckland soils, which precluded the arable agriculture practices that destroyed them elsewhere. During prehistoric times Breckland may have been one of the most densely populated regions of England, but at the time of the Domesday Book there were perhaps only half the number of people per square mile compared to elsewhere in the county. A long history of village desertion stretches back to the Iron Age and earlier, through the Roman and Saxon periods, though most instances date to the medieval. With 150 or more, Norfolk has more deserted villages than almost any other county in England, and Breckland has the highest concentration. While coastal settlements were typically destroyed by erosion, sand drifting caused the desertion of inland villages, notably including the now-reconstructed West Stow near Bury St Edmunds, which was abandoned in the middle of the first century CE when sand inundated the entire settlement and surrounding agricultural fields to depths of over a metre in places. The Black Death – bubonic plague – in England in 1348 subsequently drove rural population decline, and Norfolk was particularly hard hit. Migration to more favoured locations was often forced by poor soils and soil exhaustion: other causes included enclosure; 'engrossment', as landowners purchased local land to assemble larger and more profitable units; and emparking, as fashionable designers such as Capability Brown and Humphry Repton laid out landscaped parks around country mansions, particularly numerous in Breckland and sometimes necessitating the removal of whole villages.

LAND-USE CHANGES

In the 1920s, the governmental Forestry Commission began acquiring land for afforestation, notably upon the higher land between the river valleys then under extensive tracts of bare ground and heath. National stocks of timber were low after the First World War, agriculture was depressed, and several large shooting estates were ready to sell land in the face of death duties and the loss of their sons in the First World War, so the Commission was able to buy Breckland land cheaply. Thetford Forest was planted from 1922 largely on open ground, representing one of the largest ever land-use change projects in England. As pines were planted en masse much of the open landscape and most of the dunes were destroyed, along with habitats and wildlife. A programme of rabbit eradication commenced, since they can cause considerable damage to young trees. Today comprising some 18,730 ha, the largest lowland plantation forest in England, Thetford is predominantly a pine forest, initially planted with Scots Pine but dominated now by Corsican Pine. This Mediterranean species is not a major plantation tree nationally but is particularly well suited to the poor, light soils of Breckland. It grows fast, tolerates heat and drought, is windfirm, more resistant to fungi and insect attack than Scots Pine, and produces good-quality timber. However, where dominant it is detrimental to ground flora and invertebrates – butterflies and moths in particular – because of its heavy needle litter and the dense shade cast compared to Scots Pine. Around 12% of Thetford Forest is broadleaved, but the oak grows slowly compared to pines and is considered uneconomic.

In common with projects in other European desert locations, work camps – 'labour colonies' – were established in the 1930s to prepare land for Forestry Commission plantations. Camps built at Breckland locations including West Tofts, Cranwich, Weeting and Brandon High Lodge housed men doing heavy manual work including scrub clearance, chopping down trees, grubbing up roots, digging ditches, quarrying and preparing forest roads. Formally known as Instructional Centres, the projects were run for the unemployed by ex-soldiers, and their main purpose seems to have been to 'restore and maintain the fitness for work [...] of men who have suffered from prolonged unemployment, and to prepare them in suitable cases for further training for industry'.[11] Around 200,000 young men entered one of 30 camps established by the Ministry of Labour across the UK. While the 1934 Unemployment Act made attendance compulsory it was never implemented, but in practice there was an element of compulsion as officials coerced individuals to attend by threatening to withdraw welfare payments if they refused to work, leading contemporary communists to denounce them as slave camps. The men worked under strict discipline, in often harsh conditions and in remote rural locations thought suitable in separating them from the 'depressing atmosphere of the coalfields',[12] and pubs and

---

11 Field, J. (2009) Able Bodies: Work camps and the training of the unemployed in Britain before 1939. Paper given at conference *The Significance of the Historical Perspective in Adult Education Research*, University of Cambridge Institute of Continuing Education, 6 July 2009. http://downloads.bbc.co.uk/rmhttp/radio4/making-history/able-bodies.pdf
12 Field, J. (2012) An Anti-Urban Education? Work camps and ideals of the land in Interwar Britain. *Rural History* 23(2): 213–228.

drink. However, post-training job placement rates were low, the schemes were judged a failure, and the camps were closed in 1939.

Extensive areas of Breckland have been reserved for military purposes, as has happened on numerous other European deserts. Such open and sparsely or unpopulated tracts offer ample opportunity to muster and train large numbers of troops and rip the ground up with tanks and explosives. Military exercises took place across Breckland from 1906 and army camps were established, one in Thetford holding 40 officers and 5,000 men in the First World War. The world's first-ever tanks and tank crews trained at Elveden in 1916, said to be the most secret place in the world at the time for the military wanted the existence of the so-called landships to be kept secret until their first deployment on the battlefields of France and Belgium. Around 600 village inhabitants were evicted, the area was patrolled by soldiers who were not fully informed about exactly what it was they were guarding, and if warreners or keepers or indeed the public should be unfortunate enough to stray into even the outer fringes of the training area they would be shot. The first military airfield was built on tracts of razed heath in 1914, and from 1936 work began to construct several large airfields in Breckland. The Stanford Training Area – or STANTA – was established in 1942 as part of a national strategy of training for the invasion of Europe. Today referred to locally as the Battle Area, STANTA comprises about 12,000 ha of heath, wetland, woodland and farmland, including at about 2,500 ha the largest extent of remaining Breckland grass-heath habitat. The training area is important for biodiversity, with numerous scarce flowering plants, butterflies, moths and mammals, and several tracts have been included in the Breckland SPA/SAC nature reserve. Troops have trained here for every major conflict since World War II, the area rings to the sound of firearms and high explosives, and of course there is no public access. The Ministry of Defence undertakes some biodiversity management alongside estate management, where compatible with military requirements, and physical disturbance including troop and vehicle activity and the use of high explosives may perversely be of benefit to conservation by creating patches of bare ground suitable for many species. Large operational airfields remain at Lakenheath and Mildenhall.

Today, the Breckland landscape has been described as being 'oddly empty.'[13] Not many people live here, and few outsiders are familiar with the region: Breckland is a rural English backwater. But in the 1950s remote Thetford, the capital of Breckland, was assessed to be in serious economic decline, and as part of a national strategy to decentralise population and industry Londoners were moved out of the capital en masse to be rehoused here in what came to be an expanded or overspill town. From a population of 4,750 in 1957, small-town Thetford underwent radical change as thousands of homes were constructed for former urban-dwellers moving from whole districts demolished in east London to new houses in the middle of nowhere – the European deserts of East Anglia. By the late 1980s the population of Thetford was 21,000, a rate of increase unmatched in Norfolk or indeed the whole country, and is now 27,000 including migrants from Russia, Poland, Slovakia, Spain and Portugal drawn by the availability of agricultural employment.

---

13 Shielsflynn (undat.) *Brecks' Special Qualities: An analysis of identity and sense of place.* Cambridge Studio: Cambridge, at http://www.breakingnewground.org.uk/assets/LCAP/Brecks-Special-Qualities-Report-low-res.pdf

PERSPECTIVES FROM THE ARTS AND SOCIOLOGY

The deserts of Europe have not figured prominently in the arts, but a few British painters and writers have drawn upon their imagery. Landscape painter John Constable is best-known for his early- and mid-period paintings of rural England scenes, realistic though romanticised depictions of green-hued and productive lowland landscapes celebrating the farmed countryside with plenty of light and colour. Many were set in East Anglia where he spent his childhood, in pictures like *Stour Valley and Dedham Church* (1814) and *The Hay Wain* (1821). But during his later period he painted numerous views of Hampstead Heath – then just outside London – works which are less eulogised and contrast markedly with his earlier verdant landscapes. Constable's Hampstead Heath is an undulating sandy plain, open, bare and depicted with a much narrower colour palette than his earlier works, emphasising darker tones and frequently set under dramatic and even angry skies. He obviously had great affection for the heath and all its moods, and one also senses that not only did he recognise the otherness of these places but knew that they were disappearing through enclosure and technological advances in farming, having witnessed the enclosure of similar instances in East Anglia. Despite the fact that by that time England's common land was frequently viewed as unproductive, unsightly and outmoded, as enclosure proceeded in the early eighteenth century he and Turner and other Romantic-era British landscape painters resolutely continued to depict these rough, unfashionable places.

John Clare (1793–1864), known as the Peasant Poet, wrote numerous poems about the natural world in which heaths figured frequently. His was a naturalist perspective upon the beauty of the English natural world and rural life, expressed in simple and direct language. In one of his most admired works, 'Emmonsail's Heath in Winter', he celebrated a former tract of Northamptonshire wilderness near his childhood home in Helpston (now Cambridgeshire) where he once tended flocks of sheep and geese, writing of bracken, gorse, heather and birds on this 'frozen plain', and the itinerant peoples of all kinds who lived there. In another poem 'The Mores' he eulogised the open plains:

> Unbounded freedom ruled the wandering scene,
> Nor fence of ownership crept in between...[14]

In a song 'Swamps of Wild Rush-Beds', Clare eulogises the heathland commons:

> Ye commons left free in the rude rags of nature,
> Ye brown heaths be-clothed in furze as ye be,
> My wild eye in rapture adores every feature,
> Ye are as dear as this heart in my bosom to me.[15]

Railing against the enclosures that he witnessed with along with the associated hedge planting, Clare bemoaned that:

---

14 Powell, D. (2009) *First Publications of John Clare's Poems*. Research Papers on John Clare, number 1. Missoula: The John Clare Society of North America, 2nd edition.
15 Clare, J. and Symons, A. (1908) *Poems by John Clare*. London: Henry Frowde.

> Inclosure came and trampled on the grave,
> Of labour's rights and left the poor a slave.

And also in the poem 'To a Fallen Elm':

> The rabbit had not where to make his den,
> And labour's only cow was drove away.[16]

William Shakespeare set some of his most dramatic scenes on British heaths. The rural places he described were not always depicted as pastoral idylls but sometimes as mysterious, inhospitable, godless and even hazardous. Macbeth's stark encounter with the three evil witches was set in a threatening wilderness, a 'blasted heath' in a violent storm with thunder and lightning rumbling overhead.[17] The natural world in *King Lear* is depicted as dangerous, the king plodding across a deserted heath in howling winds and battered by heavy rain. In both tragedies, the British deserts serve as desolate places outside the civilised realm, providing backdrops for intense emotions. The 1878 novel *The Return of the Native* authored by English novelist Thomas Hardy is set upon 'Egdon Heath', a fictional location inspired by several real former heaths which he knew located in the county of Dorset in south-west England. The role of the heath was to set the atmosphere of the ultimately tragic novel – dark and sombre. Hardy paints the heath as a setting for extreme weather, intense emotions and as an arena of death, with Eustacia dying there on a dark and stormy night and Mrs Yeobright bitten by an Adder on an intensely summer hot day.

Concepts of the sacred and the profane offer interesting insights into historic social attitudes and values towards the British European deserts. Were the European desert wastes not profane places? Durkheim proposed the division of the world into two domains, one containing all that is sacred and the other all that is profane. Sacred spaces are those deemed to be hallowed locations with special significance and a connection to the divine, while the profane can be anything or anywhere that is not sacred. The differentiation of places into the sacred and the profane is as old as humankind. Historically, and in a society where the divine was omnipresent, the terms sacred and profane were linked to social boundaries and distinctions, and by extension to geographical locations. Applying the sociology of space, Breckland and similar desertic wildernesses may be characterised as profane locations in the eyes of some. Regarded as displeasing to the eye, low status, perilous and even unholy, they were secular and unconsecrated places available for profane uses unacceptable elsewhere. For Foucault, there existed in the Middle Ages a hierarchic ensemble of places: sacred places and profane places, with the latter frequently described in terms indicating that they were stigmatised and regarded as marginal – at least by some – and ungodly. Recurring themes in the limited available historic literature condemn the British deserts for their lack of productivity, appearance, inhabitants – and their bad reputations and role as a locus for criminality and profane

---

16 Clare, J., Blunden, E. and Porter, A. (1920) *John Clare: Poems chiefly from manuscript*. London: Richard Comden-Sanderson.
17 Braunmuller, A.R. (ed.) (1997) *Macbeth: The New Cambridge Shakespeare*. Cambridge: Cambridge University Press.

activities. For some, they were the risky 'no-go' zones of their respective rural regions, in the same way that today some urban neighbourhoods might be avoided because they are deemed undesirable or even hazardous. The British European desert wastes were the setting for activities and people considered worldly, unlawful, blasphemous and heathen.

When in 1615 John Norden reported on the English heathlands, the attitudes of at least some in society towards their inhabitants were expressed in a clear and undiplomatic fashion. The English cartographer and surveyor of the Crown Lands recorded that everywhere in the then-extensive common wastes there was:

> Idleness, beggary and atheism [...] wherein infinite poor, yet most idle inhabitants have thrust themselves, living covertly without law or religion [...] among whom are nourished and bred infinite idle fry, that coming to ripe grow vagabonds and infect the commonwealth with a most dangerous leprosy.[18]

There are numerous references to criminality associated with the English desertic regions. The small town of Brandon in central Breckland seems to have had an unusually large number of inns, none of which enjoyed a particularly salubrious reputation and one apparently attracting 'lecherous and suspicious' men.[19] men. Both Brandon and Thetford were reputed to be favoured haunts for criminals and fugitives from justice in medieval times, and the 'honest men' of Thetford were scandalised when in 1467 local Andrew Henke condemned the town's lawlessness and pronounced it full of thieves. The surrounding heaths offered tempting quarry for poachers, and innkeepers were among the leaders of poaching gangs. The courts here and elsewhere in the fifteenth century attempted to regulate social behaviour, and sentences for the convicted were harsh. In 1813 one man was sentenced to be transported for seven years for taking a single coney from a warren; and in the last executions in Thetford in 1824, three men were hanged for sheep stealing. Deer poachers still operate in twenty-first-century Breckland, working with night vision equipment and high-powered rifles.

BAD LANDS, BAD PEOPLE

A notorious instance of a European desert historically regarded as a profane place is Tiptree Heath, 80 km south of Breckland. Located on a ridge of glacial sand and gravel, Tiptree Heath once straggled across thousands of hectares of north-east Essex. Best known today for the nearby Wilkin and Sons company growing soft fruit to produce 'Tiptree' brand preserves, the heath is common land first recorded in 1401. For centuries the tract served as communal rough grazing for domestic livestock, and was extensively quarried for sand and gravel and used as a rural rubbish dump. With a bad reputation as a locality and a name for lawlessness, Tiptree Heath was described in 1850 as:

---

18 St. John, J. (1787) *Observations on the Land Revenue of the Crown.* London: J. Debrett, p. 168, Appendix II.
19 Bailey, M. (1989) *A Marginal Economy? East Anglian Breckland in the Later Middle Ages.* Cambridge: Cambridge University Press.

> A waste and wild country covered with furze [...] and extremely bleak [...] Here are all the indications of common rights and poverty, irregular and decayed buildings and numerous but scattered population, plenty of geese, donkeys, bad fences, and beer shops, a few windmills, and a few crossroads and sign posts. Tiptree has a very bad name [...] and a common saying in Essex, on hearing it mentioned, is: 'Tiptree Heath! God help you!' An Irish gentleman who visited said of the neighbourhood: 'I have seen nothing in all Ireland so bad.'[20]

Close to the secluded creeks of the Blackwater estuary, the heath was known as a haunt of smugglers who shipped contraband goods clandestinely to avoid paying customs duties. Wines, spirits, tobacco and silk brought across the North Sea were transported there on horses and donkeys, to be concealed in shallow holes dug in the sandy soil and covered with turf and brushwood or stashed in safe houses prior to sale. Large flagons of wine and spirits were submerged in a local pond.

Similar to numerous other British deserts, Tiptree Heath was said to have been populated by itinerant peoples of all kinds: hawkers, tinkers, squatters and 'landless men', vagrants, footpads, fugitives from justice, numerous unemployed,[21] and a witch of the name Moss who lived on Bull Lane. Throughout the UK the desertic heaths with their temptations of unenclosed land attracted travelling paupers, who were said to seek out parishes with 'the largest commons or wastes to build cottages, and the most woods for them to burn and destroy and when they have consumed it then to another parish'.[22] Some individuals may have resorted to living on or beside heaths after having been evicted as a result of enclosures. Squatter settlements of chaotically scattered cottages were constructed with whatever materials were available locally, including what was called mud: in fact clay, sand, straw, water and earth on a timber-and-branches frame. Called cob houses, they were linked by narrow lanes twisting between the irregular smallholdings, such 'encroachments' frequently cutting across the rights of local farmers and landowners. An unusually positive depiction of these heath people was essayed by the artist and natural history author Thomas Bewick in his 1862 memoir. Describing the cottages – 'hovels' – scattered across the Northumberland heaths, which had been built mainly by their inhabitants' own hands and at their own expense, he observed:

> These hardy inhabitants of the fells and wastes, whose cottages were surrounded with whins [gorse] and heather, I must observe that they always appeared to me, notwithstanding their apparent poverty, to enjoy health and happiness in a degree surpassing that of most other men [...] When tired, at night, with labour, having few cares to perplex them, they

---

20 Bamford, E. (1996) *Tiptree: One Day a City? The Story of Tiptree Part One*. Elaine Bamford, Owl Printing.
21 Boot, F. and Davenport. A. (1977) *The Creation of a Village: The Story Of Tiptree*. Tiptree: Workers' Educational Association.
22 Ward, C. (2002) *Cotters and Squatters: Housing's Hidden History*. Nottingham: Five Leaves.

lay down and slept soundly, and arose refreshed from their hard beds early in the morning.[23]

For centuries fairs were held on Tiptree Heath annually in July. In the late eighteenth century this popular rural revel was said to have lasted for a month, but was later restricted by magistrates to a two-day duration. Attractions included various performances; amusements for the children; coconut shies; stalls selling drinks, fruit, vegetables, cockles and whelks; and horse racing, though this ostensibly central reason for the event's existence was often regarded as subordinate to it being simply an opportunity for a bucolic social revel. The 1831 edition was marked by a brutal boxing match which culminated in the death of one contestant and the arrest of the other. The Tiptree Heath fair and races became less frequent after the First World War, but continued sporadically into the 1920s and after. The remaining fragment of heath is now a nature reserve. Such illicit or at least frowned-upon bucolic merrymaking events were routinely held on the English heaths, these profane spaces imbued with a subversive role and where riotous scenes of shocking debauch offering a threat to the wider moral order were traditionally tolerated. At a similar annual 'frolic' in Wenhaston in eighteenth century east Suffolk – still a heath village today:

> Noisy merriment reigned supreme and there was much betting on pony and donkey races that had as their goals the two public houses, the Queen's Head at Blyford and The Star at Wenhaston. Masters were as degraded on these occasions as their men; indeed worse, for they were naturally able to keep up their revels longer. Some farmers would drink day and night for sometimes a week without going home, one after another paying in turn rounds or reckonings reaching £2 per round.[24]

Elsewhere the Fairs Act of 1871 had been employed to abolish numerous similar traditional fairs held in early Victorian England because they were deemed unnecessary; the cause of grievous immorality; and very injurious to the inhabitants of the towns in which such fairs are held. Long-established London fairs were quashed, but some held on the city's outskirts continued to offer all the fun of the fair, including those on Hampstead Heath and Blackheath. Hounslow and other heaths were favoured locations for duels, even though killing someone in a duel was illegal; and cock fighting was thought to have been held on Nomansland Common in Hertfordshire until the early twentieth century, despite having been illegal in England since 1835. Funfairs continue to be held on commons in the twenty-first century, including for instance that at Melford Common on the edge of Thetford.

Bad places, bad people: profane lands, profane people? There are plenty of historic links between the European deserts and crime, criminality and punishment. Public humiliation was once an integral part of the justice system, and punishments offering

---

23 Bewick, T. and Bewick, J. (1862) *A Memoir of Thomas Bewick*. London: Longman, Green, Longman, and Roberts.
24 Clare, J.B. (1903) *Curious Parish Records and Workhouse Riots*. Halesworth: William P. Gale.

what was termed a public spectacle of horror provided entertainment, signalled a warning to others, and symbolised the executed criminal's exclusion from normal society. A man of the name Watson was executed by hanging in Thetford in 1795 for the murder of his wife, Elizabeth. It was the custom at the time to display murderers' bodies in public by suspending the corpse in a human-shaped iron cage in a practice known as gibbeting. Where might have been chosen as the location for Watson's corpse to be displayed? Bradenham Common, once a heath on the north-east margins of Breckland, where the murderer's body was said to have hung for many years. A century later, the remains of the gibbet cage were unearthed on the common by the novelist H. Rider Haggard in 1899 with a piece of human skull still attached to it, thought to have been Watson's. Gibbeting was abolished in 1834; Bradenham was the last gibbet to stand in Norfolk; and the surviving gibbet cage is in the care of the Norfolk Museums Service complete with the fragment of skull. Gibbets were often located on the profane British deserts, the bodies consigned to these places outside consecrated ground rather than being honoured with a Christian churchyard burial as was customary.

Thetford Heath grassland with heather and gorse.

About 170 km further south from Bradenham, another European desert at Hounslow Heath was once described as the most dangerous place in England. Formerly on the outskirts of London and large at 1,600 ha (80 ha today), the heath was described by William Cobbett in 1830 as 'A sample of all that is bad in soil and villainous in look.'[25] Across its wilds ran the main routes from London to Portsmouth, Bath and the west of England. Hounslow was the scene of hundreds of hold-ups, with mounted highwaymen stealing from stagecoach travellers on the isolated and secluded heathland roads. Once apprehended and executed – perhaps at Tyburn – highwaymen's bodies were carted off to rot on the heath *pour encourager les autres*. So plentiful were the gruesome gibbets at Hounslow that some were regarded as landscape features and marked on eighteenth-century maps.

HEATH HAZARDS

Many European deserts are hazardous places to be even today. Wildfires have long been a threat at these parched locations. Fire destroyed 17 ha of Canford Heath in Poole in April 2022, one of the largest remaining areas of heathland in Dorset and a designated nature reserve. There was an intense blaze, with thick plumes of smoke and huge sheets of flame driven by strong winds and unusually dry vegetation. Twenty households were evacuated to a nearby leisure centre as the fire approached to within a few metres of their homes. The fire service stated that a degree of human intervention may have been involved.

Some of the wildlife of these places poses a threat to humans. Breckland is one of the strongholds of the Adder, as are other British deserts, the only venomous species of the three native British snakes. Growing up to 60–80 cm in length and with a distinctive black zigzag pattern along the length of its body and piercing-red eyes, the handsome Adder – also known as the viper – is easily identified. With a preference for open areas of warm, light, sandy and well-drained soil with scrub and open woodland nearby for sheltering from predators, heathland is ideal territory for snakes, and the secretive Adder may be glimpsed basking in the sun and foraging for small mammals, amphibians, lizards, birds and insects. While non-aggressive, the snake's presence charges the landscape with caution simply because it is capable of killing humans. Attacks are rare, estimates indicating 50–100 Adder bites to people each year in the UK. The snake usually only attacks if disturbed or provoked, or the creature is picked up. While most bite victims experience only a minor reaction, medical attention should be sought immediately, though usually only the very young, old, or ill are in any real danger. Bites can be painful and the venom can produce severe effects including kidney failure, anaphylactic reactions, heavy blood loss, coma and even cardiac arrest. Fatalities are rare, with 14 recorded human deaths since 1876 – the last in 1975 – indeed, deaths from bee and wasp stings are more common. But for instance an Adder on Hounslow Heath slithered into a two-year-old boy's pushchair in April 2017, and as the father reacted by attempting to knock the snake away it struck, biting his right hand:

---

25 Cobbett, W. (1893) *Rural Rides Vol. 1*. London: Reeves and Turner.

European deserts hazards – the aftermath of a wildfire on Rushmere Heath, Suffolk.

> Within two minutes I was on the floor. I couldn't move, I was throwing up straight away. I think everyone was a bit worried [...] I've never felt anything like it – my whole body was in agony. I was paralysed and couldn't talk to anyone.[26]

Foaming from the mouth, the father was rushed to West Middlesex Hospital. There, his heart rate was said to have decreased to about four beats per minute, but after antivenom had been administered he was able to leave hospital. Despite being a protected species in the UK, Adder populations are in decline throughout the country – notably including intensively farmed areas like East Anglia – and there is concern that, if current trends continue, the snake will be restricted to a few large population sites within 15–20 years, significantly increasing the extinction risk for this UK priority species. Reasons for its decline include fragmentation and loss of its heathland habitat, 'public pressure' and habitat disturbance, especially by dogs, and perhaps too the Pheasant, which will attack an adult Adder by pecking and swallow young individuals whole; its dense feathers are thought to protect it from the snake's bites. Over a century ago, a British man called Harry 'Brusher' Mills achieved widespread celebrity for his self-appointed

---

26 Mitchell, J. (2017) Father left fighting for life after being bitten by adder in west London park. *The Standard* newspaper (London), 12 April 2017.

profession as snake catcher, trapping Adders and Grass Snakes with a forked stick in the New Forest, another prize British desertic tract and National Park in southern Hampshire. Those he caught were usually sent to London Zoo as food for snake-eating birds of prey.

Another hazardous heathland creature is very small indeed. Ticks are blood-sucking parasites feeding on birds and mammals, and have an affinity for humans. The key species is the Sheep, Deer or Castor Bean Tick, a tiny 3 mm long spider-like creature found all over the country but particularly in grassy heath and woodland areas where wild deer and rabbit are in abundance, notably including Breckland. A single infected bite from this innocuous-looking bug can cause Lyme disease, initiating an unpredictable and wide-ranging spectrum of symptoms ranging from the mildly unpleasant to the severely debilitating. With an ambush strategy, the tick ascends grasses, twigs and branches where it waits for a host to brush by – at which point it clings on and its mouthparts penetrate the skin to feed, though the unfortunate host will feel nothing. The tick subsequently detaches to lay its eggs and die. *Borrelia* bacterial pathogens may be transferred via the tick's saliva, causing Lyme disease. If untreated, the infection may produce fever, rash, severe headaches, fatigue and muscle and joint aches. Later stages may bring severe neurological conditions including facial paralysis, severe joint pain and swelling, heart irregularities, inflammation of the brain and spinal cord, and problems with memory or concentration. People have gone blind. No more than 4% of ticks may be infected, and early treatment with antibiotics is usually successful. But no vaccine is available, and as control is impractical, avoidance and preventative measures are crucial. In the UK, tick

The Adder, or European Viper, is usually associated with open heathland, and is the only venomous species of the three native British snakes. (Bernard Dupont, CC BY-SA 2.0)

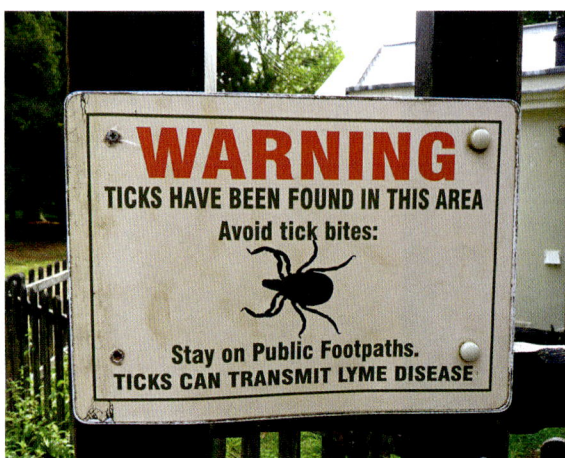

Hazards – sheep tick warning sign near Aspley Heath, Bedfordshire.

numbers are increasing year on year, people are at risk throughout the country, and as awareness of the risks posed by this tiny creature grows warning signs have been erected at some countryside locations. Between 1,000 to 3,000 people are diagnosed with Lyme disease each year in England, though the true figure for infections may be three times greater.

BRECKLAND IN THE TWENTY-FIRST CENTURY

In common with many semi-natural areas in the UK, the Breckland habitat is under threat. Today's remaining desertic Breckland tracts are the remnants of formerly extensive expanses across the region, and some are in poor condition: so much has been lost. During the twentieth century, major changes have occurred with bare ground, annuals and lichens being replaced by perennials, rank grasses, scrub and woodland. Changes in land use, the decline of the rabbit, and a general reduction in physical disturbance have fragmented the remaining areas of habitat, putting rare species at risk. An estimated 76% of the area of grass-heath was lost over the twentieth century as Wavy Hair-grass has spread to dominate Breckland to the exclusion of heather. The spread of this once-rare grass has been attributed to eutrophication, elevated soil nitrogen levels deriving from air pollution and probably originating from surrounding areas of intensive farming. This primary limiting nutrient enriches and thus degrades the infertile habitat via rainfall. Active management is required to maintain the open, nutrient-poor conditions required by so many Breckland species.

Lowland heathland is one of the scarcest and most threatened habitats in Europe. These unspectacular parched expanses are valuable, but vulnerable too. Centuries ago, a substantial proportion of England lay under barren tracts of waste land – perhaps as much as one-third – an extraordinary fact attested to by the fact that the word 'heath' occurs today in about 2,000 English place names, a figure which increases to 2,500 or more when Scotland and Wales are included. Lowland heathland is a critically important habitat for all 13 of the UK's reptile and amphibian species and half of British dragonfly

UNITED KINGDOM: BRECKLAND – AN ARID DESERT IN VERDANT EAST ANGLIA

Thetford heaths landscape.

The Common Slow Worm is a species of legless lizard native to western Eurasia that can grow to a length of 50 cm and superficially resembles a snake. It is a semifossorial (burrowing) species, spending much of their time hiding underground or beneath objects. (Agnieszka Kwiecień, Nova, CC BY-SA 4.0)

Cavenham Heath, Breckland. (Angela Fraser)

and damselfly species. Yet 84% was lost between 1800 and the late 1980s, largely due to the impacts of agricultural reclamation, afforestation and urban development. Today Britain has a significant proportion – 20% – of the European lowland heathland total and therefore special responsibility for its conservation. The habitat is scattered throughout the UK, though mostly concentrated in the southern and eastern counties of Cornwall, Devon, Dorset, Hampshire, Sussex, Surrey, Suffolk and Norfolk.

Of high nature conservation importance, Breckland is recognised as significant on an international scale. Twenty-five species previously recorded in Breckland are now considered to be nationally extinct, ranging from flowers and mosses to beetles and moths. The Silver-studded Blue, a handsome butterfly with distinct metallic spots on the underside hindwing, is mostly restricted to heathland with heather and gorse nectar sources. This small blue butterfly has disappeared from Breckland, a former stronghold, and is scarce and declining over the UK as a whole though still present on the Sandling heaths of east Suffolk. One endangered plant restricted to Breckland is the Prostrate Perennial Knawel. Found here alone, the species is adapted to the sandy, nutrient-poor soils of Breckland but outcompeted by other more vigorous plants. Its distribution has declined over the past century from 36 sites to three, one of which – with just a few hundred individuals – represents about 95% of the world population. The extent of the loss of the UK's biodiversity over the last 50 years, including plants, animals and fungi species, has seen the country labelled as one of the most nature depleted. There have been significant losses of terrestrial plants adapted to low fertility conditions and low competition, the result of habitat loss linked to intensive agricultural practices, afforestation, development and air pollution. In 2022 the UK made a formal commitment at the UN Biodiversity Conference to halt and reverse nature losses, agreeing to protect and conserve a minimum of 30% of land and sea for biodiversity by 2030 through enhancing and expanding existing protected areas and establishing new ones. However, targets for heathland should be included in national policy pronouncements.

As the Breckland environment changes, invasive sand is now mostly a historical memory rather than a twenty-first-century hazard. Part of Lakenheath, the site of one of the last working warrens which survived until 1940, remained covered in sand dunes up to 6 m high in the 1930s. A monster 1 km-wide dune there was levelled during World War II. Only fragments of this dune system remain. Interestingly enough, one of the consequences of the construction of the Lakenheath airbase in 1940 was the conservation of over 100 ha of heath, for airfield operations require an open landscape setting, and the area is now managed by the United States Air Force. More fragments of dunes remain at the nearby Wangford Warren Nature Reserve – once a large warren formerly described as a miniature Sahara – though many now lie under hectares of conifer trees.

The handsome Silver-studded Blue is principally a butterfly of heathland, where the caterpillars feed upon heathers and a variety of pea family plants. (Guy Padfield)

Other relict dunes are preserved at Foxhole Heath and Icklingham Plains among other locations where they are fixed under vegetation, but there are no longer any active dunes in Breckland. Flyaway sand remains a live issue, however, with what are termed 'fen blow' winds gusting up to 88 kph and lofting clouds of sand. Road clearance may be necessary after high springtime winds, and occasionally necessary it is for vehicles to use their headlights for better visibility during daytime sandstorms.

BRECKLAND CONSERVATION

Projects aimed at the restoration of Breckland habitat are in progress, though given the nature and scale of both the threats and the afforestation undertaken over the last century the prospects are for marginal change only. Once a widespread and familiar farmland and downland bird over much of southern England, Stone-curlew is a scarce species across Europe but has been benefiting from intensive conservation efforts in the UK, where its breeding population has grown to some 350 pairs from a low point in the mid-1980s when only about 100 pairs were breeding, two-thirds in Breckland. Stone-curlew feed on earthworms, soil-surface invertebrates, slugs and snails, which are easier to find in areas of bare and broken ground than in thick grass, so ploughing and harrowing areas of grassland creates attractive foraging and nesting sites. Rabbits produce conditions perfect for Stone-curlew, and boosting their numbers is the goal of trial projects constructing simple brush piles of felled branches, uprooted trees and bushes to encourage burrowing and provide cover against predation. Furthermore, mechanical diggers scrape away the top 5 cm of topsoil to expose plots of bare sand, piling it up into banks of about 1 m in height to provide a suitable substrate for burrowing and breeding in much the same way that Breckland warreners did in the past. There is no little irony in the fact that efforts to expose bare sand are being undertaken in a region where drifting sand has been a historic environmental hazard! Other species benefit too, including rare early pioneer successional plants, for measures good for Stone-curlew are invariably good for others. Farming interests have expressed concerns at the prospect of increased rabbit numbers, for some regard the creature as a pest species, but many big estates with a sporting interest – rabbit shooting – hold a different view. Initial indications suggest some success with these measures, but the grim impacts of rabbit haemorrhagic disease have obscured any positive results. A perhaps more fundamental goal of restoring conifer plantations to open Breckland plains has resulted in a handful of cleared sites, some of which are now of nature reserve quality and support rare plants including Spanish Catchfly and Proliferous Pink. A major gatekeeper, Forestry England is involved in the achievement of biodiversity and landscape goals in Breckland, but timber production remains its primary raison d'être.

Driving away south through the barely differentiated arable fields of Norfolk and then Suffolk, I yearned for the Breckland landscapes that lay behind with their motley appearance and hazy, enigmatic ambience. Breckland may be marginal in conventional land-use terms but is anything but in biodiversity terms, not least because many of its soils may never have been cultivated or subjected to a cocktail of agrochemicals. The admixture of heath, steppe and coastal species here, perhaps unique at a national scale, added to the geology and impacts of cultural practices over millennia, have made Breckland a fascinating place with an extraordinary wealth of wildlife. This remote region has undergone several thoroughgoing environmental transformations over the last few millennia. Oh, to be able to peer into the past and view the pre-forestry landscape of a century and more ago; but today there are few places remaining whose appearance approximates to the sweeping rough-country wilderness vistas of centuries past.

Breckland wilderness is now beginning to be recreated, though as yet only on a limited scale. Once condemned as an exotic and invasive pest species, the rabbit is again being courted with human-managed rearing environments to foster its fortuitous prowess as a key species supporting the promotion of heathland biodiversity. While some existing wildlife sites continue to persist in unfavourable condition, there is immense potential in Breckland for much more than simply remedial conservation management. An opportunity exists to achieve substantial gains through large-scale biodiversity and land restoration strategies involving habitat restoration and recreation on a large scale, with the goals of halting the decline in species populations and addressing the issue of the fragmentation of existing sites through increasing habitat connectivity. The prospective gains in Breckland are surely proportionately much more significant per unit of resource expenditure than those achievable in most other locations in East Anglia.

Fundamentally, however, there is little sign of any *vision*, nor indeed a comprehensive plan for the UK's European deserts at the national scale. Such an ambitious – and even radical – vision is a necessary prerequisite to secure the future of these desertic tracts, one translated into quantified and substantial goals and supported by long-term leadership and commitment. Nothing less will do for the future of Breckland, and indeed for the UK's dynamic European deserts in aggregate. Without goals and targets, and responsibilities apportioned between the multiplicity of agencies and stakeholders involved, the doubtless valuable measures currently being implemented appear to amount to little more than an ad hoc approach. Breckland has not been assigned Protected Landscape status; there is an absence of any coordinating entity providing clear purpose and direction in the region; and neither is formal guidance available for Breckland land managers. The 2024 UK government targets to restore or create more than 250,000 ha of a range of wildlife habitats seem not applicable to the region, and no mention is made of heathland at all. For all sorts of reasons these East Anglian deserts are perhaps easy to disregard compared to other habitat types that are more conventionally attractive, more readily understood, and perhaps less demanding in management terms. Yet these unspectacular tracts represent one of the country's most significant contributions to European and global biodiversity, as well as offering extraordinary and historic landscapes. While the challenges involved in assuring their long-term conservation are considerable, these wildernesses merit a far higher priority than they are presently ascribed.

Chapter Eight

GERMANY: A SANDY HIKE THROUGH PURPLE HEATHER

The Heidschnuckenweg is a long-distance walking trail that is just perfect for the European deserts aficionado. The route runs for 223 km out of the city of Hamburg in northern Germany due south to Celle, a handsome small town where a gleaming white *schloss* in a wild mix of Baroque, Gothic and Renaissance styles overlooks an old town which, with over 400, has the largest assemblage of timber-framed houses in Europe. After overnighting in Hamburg's busy city centre, I took the bus to Fischbek on the south-west outskirts of Hamburg, a leafy suburb from which the Heidschnuckenweg trail commences. Here the zealous adventurer strides out almost immediately onto heathland – the north German wilderness starts here!

On a milky-white day I was heading for Lüneburg Heide, named after the nearby town of Lüneburg and about one-third of the way to Celle. No vertiginous mountain walk this: the Heidschnuckenweg is an elongated lowland tramp across sand, sand, and more sand. The sweeping German desert wastes through which the trail winds are termed *heide*, a near-enough direct translation rolling up heath, heathland and heather. There towards the end of the heather flowering season, the lands around me were adorned with purple flowers at ground level with a gentle, subtly sweet aroma. Over its course the Heidschnuckenweg is largely flat, but there are some hills along the way yielding great views. Discreet signposts guide the walker through the 'wastes' which I seemed to have all to myself, having encountered no others obviously attempting a multi-day walk. The first surprise were the anthills, several prominent, large and beautifully symmetrical dome-shaped examples located on the heath margins alongside pine and birch woodland, one nearly a full metre in height.

HEATH SHEEP AND BEES

Dotted over the rolling yellow-sand expanses were sheep. Historically, the German heath farmers engaged in *Heidebauernwirtschaft* – a German super-word translating simply as heathland farming – having developed a successful approach to making optimal productive use of the region's nutrient-poor soils. This centred upon the classic German heath sheep breed, the Heidschnucke. The name translates as something like 'moorland sheep', more formally known as the German Grey Heath. The Heidschnucke is closely

Heather in flower on Lüneburg Heath.

Anthill on the outskirts of Hamburg.

associated with the northern German heathlands, where it has for centuries been the breed of choice for *heide* farmers. Kept for its meat and wool – but also, and importantly, its nutrient-transfer capabilities – the sheep is thought to descend from the mouflon, the wild sheep of Corsica and Sardinia, whose genetic lineage traces back to the dry Caspian region of eastern Turkey, Armenia, Azerbaijan and Iran. This short-tailed ruminant variety is hardy, small and with a low body weight, and said to be relatively easy to look after. The ewes are prolific, lose little body weight despite the impact of birthing and suckling, and the lambs gain weight at a greater rate than other varieties. The Heidschnucke excels in its ability to thrive in poor conditions, linked at least partially to its intake of plants which other breeds disdain – for in addition to the ability to feed on heather they are also capable of browsing trees and shrubs, though they avoid prickly plants like juniper and blackthorn. The Heidschnucke produces a high-quality, tender and very lean meat with an intense flavour likened to that of game animals such as venison, though their long, straggly grey wool is suitable only for coarse fabrics such as carpets. They keep their distance from walkers, but seem unperturbed by their presence.

These German European deserts owe their continued existence to millennia of heath farming. Without the Heidschnucke sheep, this cultural landscape would probably no longer exist. The significance of the humble sheep has been celebrated in various ways. The Heidschnucke is the logo of the Lüneburg Heide Nature Reserve, and its name has been attached to the long-distance walking trail, Heidschnuckenweg roughly translating as 'The Way of the Sheep.' The Heidschnucke has also been granted Protected Designation of Origin status, indicating an agricultural product originating and prepared in a specific area of origin using traditional methods, the designation serving to guarantee its authenticity, characteristics and reputation. A traditional way to serve Heidschnucke is in a sweet-and-sour style with a honey and juniper berry sauce, baked in the oven or boiled and served with potatoes and beer, for this is Germany. Heidschnucke may provide some income in the form of meat and wool sales, but keeping these heritage animals is expensive.

Heath sheep the Heidschnucke – the German Grey Heath. (Quartl, CC BY-SA 3.0)

More closely related to the wolf than the jackals of Africa, the Golden Jackal is native to Eurasia. (Alpcem)

I was expecting to see sheep along the Heidschnuckweg, and maybe also cattle – but not goats. And certainly not goats in their hundreds, and seemingly unfenced: free-roaming. A so-called landscape manager species, these heathland goats or *ziege* appeared friendly, intelligent, inquisitive and very much at home on Lüneburg. Those in the herd I saw were mostly white in colour with brown patches and cute floppy ears, capable of standing on their hind legs to browse the lower branches of the few trees, and seemingly very content. They too have little fear of humans, and seem to rub along well enough with the Heidschnucke. Both creatures are at risk from the wolf, however, and the Golden Jackal, originating from south-eastern Europe but now spreading naturally to northern Europe. Goat milk, cheese and meat are available from small farms in the region.

Sheep were for centuries the predominant livestock of the German heathland farmers, but – however unlikely it may seem – a significant though secondary role was played by bees. The heather, this hardy evergreen dwarf shrub blanketing much of the *heide*, offers excellent pasture for bees, and along with bilberry and cranberry heath honey is one of the best-known products of the German deserts. For long, *honig* was the only sweetener available in Europe, and *heidehonig* (heather honey) production was an important economic sideline on the heath in the Middle Ages. Beekeeping was widespread across Lüneburg, most farms would have employed a beekeeper, and honey production was systematised to a degree. The flowers of the heather plant may be small, but there's an awful lot of them. Plants rely on insects for their fertility, with sugary floral nectar and pollen from flowers being the rewards offered to visiting insects which in return act as pollinators.

The European Blueberry or Common Bilberry (aka Whinberry or Whortleberry) produces delicious edible fruit. Native to almost every European country, it may be found in a range of acidic habitats including heaths. (Kora27, CC BY-SA 4.0)

Lüneburg's heathery expanses supply the bee colonies with nectar for energy and the production of honey; the pollen provides protein and other nutrients. As the bees go from plant to plant, they carry pollen from one to another, pollinating and fertilising the flowers as they go and making a fundamental contribution to the maintenance of the heathland flora and fauna. Heather is a valuable food source for pollinating insects, as 24% of its rich nectar consists of sugar, and the plant's extended flowering season lasts from summer through to mid-autumn, a time late in the year when few other flowers are available. A heather stand might produce in the order of 100 kg of bee-collected nectar

Bee hives on Lüneburg Heide.

per hectare over a single season, a yield which compares very well to that from fruit trees like apple, cherry and plum which produce only 10–20 kg/ha, and sunflowers at 40–50 kg/ha. However, despite its substantial yield, heather is not in the same league as the Lime tree with up to 800 kg/ha – or the beautiful Globe-thistle, no desert species but one of the most productive at up to 1,000 kg/ha.

Historically the bees were housed in skeps, woven hive baskets made of straw in the shape of a bell or a rectangle, sealed with a mixture of cow dung and peat, and moved from place to place to take full advantage of the fluctuating feeding opportunities across the landscape. Today's hive boxes are made of wood or plastic and are brought here in the nectar season to be set up on the *heide* in rows in timber shelters. Heather honey is dark amber in colour, jelly-like and has a distinctive woody, warm and fresh floral aroma reminiscent of the scent of the flowers. Honey production varies depending on several factors including the weather, with for instance poor flowering after long periods of cold and rain yielding a lesser quantity of nectar compared to that after a warm and dry summer. Experts are said to be able to differentiate one year's honey produce from another. Charmingly, the grazing Heidschnucke sheep help the honeybees' mission by sundering the spiderwebs strung between heather plants, thus enabling them to fly freely.

BIG LANDSCAPES AND HEATHER

Around 40 km south of Hamburg the path heads up to Lüneburg Heide. This is a substantial area of heath, a desertic environment formerly commonplace across the North European Plain. Lüneburg Heide comprises the core Lüneburg Heide Nature Reserve of 23,440 ha, first established in 1910 and comparable in size with Birmingham, England's second city. This strictly protected tract is set within the broader Lüneburg Heide Nature Park, Naturpark Lüneburger Heide, totalling 107,800 ha, which serves as a so-called buffer zone around the core reserve. With more than 5,000 ha of bitingly dry sandy heath, Lüneburg Heide is nothing short of majestic, a stunning instance of a European desert. Lüneburg Heide Nature Reserve today constitutes the largest remaining contiguous area of dwarf shrub – heather – heathland in central Europe. The vistas here are of extensive and seemingly boundless gently rolling sandy plains. A landscape of alluvial sand deposits, physically Lüneburg is a broad landform extending for about 90 km between Hamburg and Hannover with an average elevation of 75 m and radially drained. Today's heath was shaped by glaciations and meltwater, resulting in a landscape of moraines – debris transported by ice – glacial outwash plains of sand and silt, and a few areas of sandy loess deposits. The flat-topped hill Wilseder Berg at the centre of the reserve is the highest point on the North German Plain at 169 m, formed as part of a glacier-pushed terminal moraine. Under a climate transitional between Atlantic and continental, many of Lüneburg's desiccated tracts are cloaked with heather, with occasional patches of grass, notably in the shallow valleys. These undulating, big-vista landscapes dotted with occasional young trees and patches of scrub have been likened to steppe, seemingly having more in common with the open plains of Hungary and further east than the small-scale landscapes of central Europe.

Large tracts of bare ground also feature on Lüneburg, a rare phenomenon in verdant and fertile Europe. Dunes in the open landscape are scarce today, though some areas of plantation pine have been cleared of trees in recent years specifically to expose the relict dune landscapes upon which they were planted. The *heide* is also known for 'erratics', large boulders deposited far from their origins by glaciers, many having been transported from Scandinavia and Finland. These were long ago grouped together to form megalithic tombs, and today over 1,000 of these Bronze Age burial mounds remain. However, some were plundered for building stone, with many of the largest boulders – of 2–3 m in size or more – used in the construction of walls and quays of the port of Hamburg, and others located on areas formerly used for military exercises destroyed. Around 3,000 ha of these empty sandy lands were formerly reserved for tank training, with the result that the ground surface was ripped up to such an extent that large areas of sand dunes developed until the early 1990s when the military zone was incorporated into the reserve and ecologically restored.

The sheer scale of the historical extent of heathland in Germany is astounding. While estimates are inevitably imprecise, heathland probably attained its greatest extent in the mid-eighteenth-century following centuries of progressive deforestation, when about half of the northern German lowlands were described as barren heathland dominated by drifting sand. Central parts of Lüneburg Heide were 90% 'waste'. More of the German environment was heath than in any other European country for which estimates are available, with perhaps one million hectares, the size of the Republic of Lebanon – or more – in the nineteenth century, at a time when the adjacent countries of the Netherlands and Denmark were estimated as having 800,000 and 658,000 ha respectively. In comparison, the UK had the least at 145,000 ha (Scotland not included), some authors attributing the relatively limited extent of arid soils on the western European islands compared to that

The European Stonechat favours open heath and rough grassland with gorse, where it feeds on invertebrates, seeds and fruit before flying back to its perch. (Fernando Losada Rodríguez, CC BY-SA 4.0)

Pine and Juniper on Lüneburg Heide with heather flowering in the background.

of the North European Plain as the reason. Since then, everywhere there has been a startling decrease in the extent of heathland, of the order of 60–95% across the countries of Belgium, Denmark, France, Germany, Sweden, the Netherlands and the UK. While the mass conversion of heathland to plantation forest, arable land and urban development was considered a significant achievement at the time, today it is viewed in terms of a loss of biodiversity and landscape, and to society. Having survived into the twenty-first century, Lüneburg is multiply designated under the European NATURA 2000 framework of protected areas as both a Special Protection Area for its bird interest – including Curlew, Nightjar, Red-backed Shrike, Stonechat and Whinchat, all heathland species – and as a Special Area of Conservation for its broader range of fauna, flora and habitats.

Lüneburg Heide is located on the North European Plain, the vast continuous geomorphological region stretching west–east from eastern England and the Low Countries to Poland, Belarus, Ukraine, and beyond. A major though unlauded feature, and mostly flat, it is one of the largest uninterrupted expanses of plain on Earth. The plain was glaciated on several occasions as the Scandinavian Ice Sheet ebbed and flowed, profoundly influencing the superficial geology and relief. Much of the area is blanketed by glacial moraine deposits, large sections are underlain by glacial outwash plains, and drainage is poorly developed throughout. Intense aeolian processes operating at the margins of the once-glacial areas – the periglacial deserts or semi-deserts – produced abundant coversands which underlie many of the European deserts. These were remobilised by human actions including deforestation, burning, overgrazing and plaggen agriculture which produced

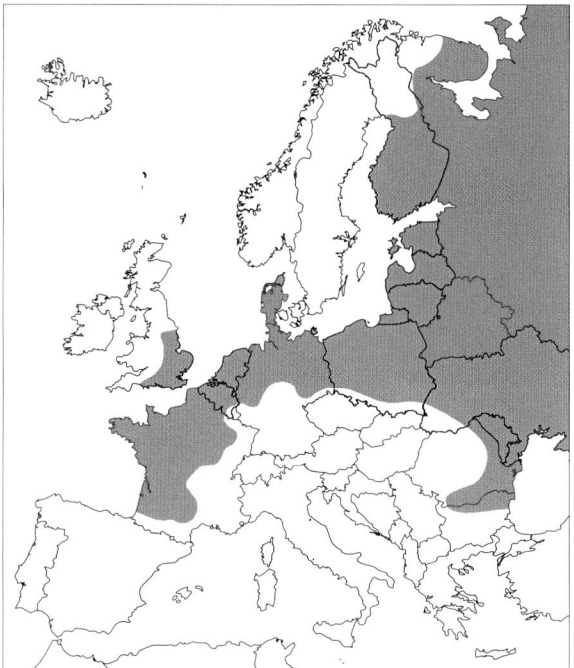

The North European Plain. (Montserrat Morillas)

large areas of drifting sand and dunes, particularly at the western extreme of the plain. Heather has dominated these lands. The eminent British heathland ecologist Professor Charles H. Gimingham (1923–2018) described Common Heather *Calluna vulgaris* as a remarkable species for its exceptional abilities including thriving in soils which are strongly acidic and lacking the mineral nutrients essential for plant growth; its capacity to dominate land which is arid and subject to large temperature fluctuations; and tolerating periods of drought. That heather is a tough plant is confirmed by its rating of H7 on the Royal Horticultural Society's hardiness scale – the highest category – for the plants are resilient even in the most severe continental European winters, at temperatures down to −20 °C.

The Heidschnuckenweg threads its way through maybe 30 or more individual heaths, large and small. It was a real pleasure to roam around this succession of deserts located at the rural heart of central western Europe. The trail offers mostly easy-going walking, and while the surroundings feel genuinely remote and do indeed resemble wilderness, civilisation is never too far distant. Inevitably, some sections go through regular countryside of arable fields and occasional woodland, affirming the reality of habitat fragmentation. Along with sweeping heath panoramas, the route passes a range of other habitats and environments typical of the Lüneburg region including shallow stream valleys, meadows, bog and wet heath. There is plenty of woodland too: about 60% of the Lüneburg Heide Nature Reserve is wooded, with forests mostly of pine planted on former areas of heath in the later nineteenth century as heathland farming became unprofitable and was gradually abandoned. At this time shifting sand dunes became more frequent, exhibiting severely

Common Heather, or Ling, is a low-growing evergreen dwarf shrub and often the dominant plant on the acidic soils of European heathland.

A shallow valley on Lüneburg Heath.

eroded landscapes, high dunes and wide-open sandy blow-outs reminiscent of the Sahara. These almost totally unvegetated expanses offered slim pickings even for sheep, and many were subsequently afforested. There also remain a few small enclaves of beech and oak forest that have survived the intensive use of past times.

LÜNEBURG VISITORS

Today the heath is a visitor destination, in marked contrast to its status a couple of centuries ago. Late summer from August to October is when the heath is at its most attractive, as the sands are blanketed with the regal lilac carpet of blooming heather for which Lüneburg is rightly famous. This genuinely spectacular sight makes a dramatic contrast to the often-featureless vistas of drab greys and browns characterising the place for most of the year. It seems clear that, without continuing management maintaining the character of its landscape, Lüneburg Heide would look very different – and probably attract fewer tourists. Hiking and horses are the favoured modes of transport here; cars and motorbikes are of course forbidden on the heath; a summertime bus service with cycle trailers connects the reserve to nearby villages; and accommodation is available in nearby villages and towns. But not everywhere is open to the public, and visitor accessibility on the nature reserve is unobtrusively managed year-round, with circular trails signposted for casual walkers while tracks crossing sensitive areas where endangered species breed are blocked off with gates and signs reading '*Kein Durchgang*' – 'No Passageway.' The Lüneburg region is now popular, attracting an estimated 35-million-day trips and 9 million overnight stays in 2019 which supported a significant proportion of the local economy. Local human-made attractions include the family theme park resort Heide Park nearby at Soltau, a large land-take attraction with vertiginous rollercoasters and Scream, a completely terrifying gyro-drop tower attracting about 1.3 million visitors annually. In 1988 the park opened a new attraction within its boundaries: Heide-Dorf (Heath Village) was conceived as a full-sized replica of the historic centre of a typical heath village, a nostalgic attraction evoking the timber-framed old town of nearby Celle. Offering a relatively tranquil refuge within an otherwise lively theme park, Heide-Dorf comprised a circular market square with building facades in the style of typical German medieval timber-framed buildings, and had a fountain, a handful of shops, a snack bar and children's amusement rides. Despite an investment of about DM 8 million, however, Heide-Dorf was unsuccessful, and was closed and fenced off at the end of the 2009 season.

Set within extensive expanses of heath, the small village of Wilsede is another Lüneburg heath village attraction. Wilsede is an authentic *Heidemuseum*, a tranquil, picture-perfect and car-free village of 40 inhabitants in the centre of the nature reserve, preserving the features and offering an insight into the living and working conditions of a typical heath farming community around 1850. The heath village receives up to 10,000 visitors each day during the high season over August and September, and the Heidschnuckenweg trail runs through here. Architectural translocation projects have brought typical *heide* farm buildings including spectacular thatched farmhouse-barns to Wilsede, all set among leafy cobblestone streets. However, there can be little doubt that the mean cottages on the heath which in former times at least some of the former *heide*

GERMANY: A SANDY HIKE THROUGH PURPLE HEATHER

Lüneburg Heide: visitors are excluded from certain areas with gates and signs reading 'Kein Durchgang' – 'No Passageway'.

inhabitants would have called home all disappeared long ago: condemned, pulled down, burned, or simply abandoned and gradually reclaimed by nature.

Like many of the European deserts, the German deserts were for centuries regarded by some as unattractive and low-status backwaters, and the people living on them disparaged too. Travellers' accounts from the middle of the eighteenth century speak of bleak, threatening and devastated desert landscapes dominated by shifting sands and offering few reasons to linger. Michel-Ange-Bernard Mangourit, adventurer and French ambassador to the United States, recounted his travels in the Lüneburg region in the early 1800s:

> We travelled over sandy plains and extensive heaths. At a great distance, geese, ducks and sheep of a very poor appearance, never failed to indicate the vicinity of some wretched hamlet. What habitations! Whole families, of the most wretched appearance, and covered with tattered garments, associate together, eat and sleep with their cattle. Near these real catacombs we observed growing a few stalks of rye and barley, and here and there a few tufts of buck-wheat. The straw is short and stunted, and the ears of a diminutive size.[1]

1 Mangourit, M.O.B. (1806) *Travels in Hanover, During the Years 1803 and 1804*. London. Printed for Richard Phillips.

EUROPEAN DESERTS

In a twenty-first century reversal of residential geography fortunes, the German heaths now not only attract visitors but also residents. Long perhaps the least-desirable places to live in their districts, localities like Lüneburg Heide are now some of the most desirable. Villages and hamlets in and around the nature reserve are now considered high-status places to live, with new residents drawn to the *heide* for its landscape, tranquillity, wide-open spaces and the contemporary social cachet associated with proximity to nature. Numerous substantial houses are tucked away along quiet tracks leading off roads heading for nearby towns and cities, and plots of land for residential building are in demand. What an ironic change in the desirability of these historically disdained places.

In common with many European deserts, few significant historic events seem to have been set on those in Germany. But one event of interest occurred at Lüneburg when, on 4 May 1945, part of the German armed forces surrendered to the British. Field Marshal Bernard Montgomery, commander of the British-Canadian forces in northern

Lüneburg Heide: the Totengrund valley with numerous Juniper shrubs, near the village of Wilsede. (Mars 2002, CC BY-SA 3.0)

Europe, took the surrender of the German armed forces in north-west Europe in a large military tent erected on Timeloberg Hill, at 78 m a low hill set on heathland 5 km south of Lüneburg town. The choice of location was said to have been symbolic: the Instrument of Surrender was to be signed within view of a defeated German city, and from Timeloberg the church towers of Lüneburg – captured by the British one month earlier – could be seen on the horizon. Montgomery later called Timeloberg Victory Hill; today it is a closed military area with no public access.

Perceptions of the German heathlands have changed fundamentally since the late eighteenth century. Jean André de Luc, a Swiss-British geologist and meteorologist, visited the area in 1776 and, unlike many of his predecessors, interpreted the heath in positive terms. De Luc was perhaps the first to describe the landscape as beautiful, and glorified the denizens of Lüneburg who cultivated the dusty sands there as noble savages in the Rousseau sense. He found them contented and happy, energised by nature and often also by the challenges it posed in their everyday lives. The noble savage literary concept held that primitive man, while uncivilised, evinced the innate goodness and virtues of those who had not been exposed to the perverting influences of civilisation – so, perhaps, harking back to the first humans: Adam and Eve in paradise. In this way the inhabitants of the German deserts were likened to the peoples in the south Pacific being 'discovered' and equally sentimentalised at much the same time by James Cook, Johann Reinhold and Johann Georg Forster. The idealised concept of the noble savage was a dominant theme in the Romantic literature of the eighteenth and nineteenth centuries, and his was an arguably overly romanticised interpretation of the people and their hard lives on the heath. De Luc also misinterpreted the origins and history of the heath itself, which he considered to be an example of virgin nature rather than the semi-natural cultural landscape of reality, and for a while the north-west European heathlands came to be misunderstood as primarily the result of nature rather than nurture, overlooking humanity's central role in the formation and maintenance of these semi-natural tracts. With the landscape no longer regarded as disreputable, De Luc's contemporaries embraced this new perspective on Lüneburg Heide, and the Romantic conceptualisation came to dominate. In 1789 the Danish scholar Jens Baggesen wrote of his delight at travelling across heaths which had previously been regarded as empty wastes:

> It has been one of my heart's desires to walk through a desert since childhood. Such an area without heights, therefore also without valleys, without wild or tame deciduous plants, without lakes, without streams, without signs of development, can be compared to a tome consisting of nothing but blank pages. [...] The further I travelled in my desert, the more pleasant and interesting it became.[2]

Over time, opinions and sentiments about the heath environment were transformed. Other creatives were inspired by the Romantic landscape of Lüneburg Heide, and as well

---

2 Baggesen, J. (1986) *Das Labyrinth oder Reise durch Deutschland in die Schweiz 1789*. Munich: Beck.

as recognising that the heath was more than simply an economically worthless sandy desert, they raised awareness of the need for its protection. Hermann Löns (1866–1914), a German journalist, writer and conservationist, became known as The Poet of the Heath for his writings celebrating the people and landscape of the north German *heide* and particularly Lüneburg. He wrote a well-known song about the heath which begins: 'On Lüneburg Heide in that beautiful land...' Visitors began to be attracted to the area which had come to be regarded as the epitome of beautiful nature, the landscape's seclusion and the seemingly natural values exhibited on these wastes offering a welcome contrast to the increasingly urbanised environments of the industrial age. By the early twentieth century, masses of day trippers sought to gaze over the miracle of the heath, their visits enabled by the arrival of the first railways in the Lüneburg area. But in the world outside the romantically conceptualised one, attempts to establish Lüneburg as the first German national park foundered when it became clear that its nature as a largely cultural landscape did not satisfy the criteria to qualify as a true wilderness – though in 1910, a *Naturschutzpark* was founded there which became the basis of today's protected area.

If Löns is probably the most famous heath poet, Günther Weißflog (1909–1987) is probably its best-known painter. Setting up house at Benninghöfen, a quiet town in the middle of Lüneburg Heide, he worked as a watercolour artist, painting the heath in a romanticised and impressionistic style. Rather than employing a muted-brown naturalistic heath palette like Constable, Weißflog's artistic interpretation of the landscape was bright, rolling, vibrant and alive with colour, bringing the charm of the *heide* to life.

Lüneburg Heide with Juniper.

Lüneburg Heide: a tract of barely vegetated ground.

He achieved a fair level of success, exhibiting in several cities, but the rural Benninghöfen cottage burned to the ground in 1955, taking many of his works with it. Moving to nearby Bispingen, Weißflog started over, but was out of necessity forced to earn a living on construction sites in Hamburg for years before in the mid-1960s he was once again able to work as an artist. Today many of his works are kept in the Bomann Museum in Celle – and four are reproduced on a visitor interpretation board located on the heath.

The trail continued its dry and dusty way through serene purple plains, passing occasional dried-up ponds; a helpful website lists the places along the trail where drinking water may be had. Though stretches of this regional trail are shared with the E1 European long-distance hiking trail that runs from Norway to Sicily, there were no other walkers to be seen most of the time. Few people live here in the north-centre of Europe's most populous country, and the general absence of people contributes to a peaceful atmosphere. The designation of the heath as a nature reserve a century ago served to restrict residential development, and few roads cross it. However, extraction of water from depths of 50–200 m to supply the taps of Hamburg has depleted the groundwater in the area north of the nature reserve. Extraction by farmers during dry periods may also have contributed to a declining water table in places, the desiccation of the ground resulting in negative impacts on vulnerable vegetation types.

BROOM AND JUNIPER

Of all the flowering heathland species, Common Broom is a personal favourite and a frequent sight in many European deserts. Widespread in Germany, where it is named *Besenginster*, this hardy perennial deciduous pea-family shrub grows erect to 1.5–3 m or more, with long, wiry, whip-like dark green stems. Common Broom produces the most spectacular flowers on the heath over the months of May to June, annually heralding the start of summer. Flamboyant clusters of large bright butter-yellow flowers explode outwards in glorious sprays from this more usually unspectacular plant, accompanied by a faintly vanilla-like aroma. The elongated two-lobed seed pods are hairy, and when mature, sun-dried and black in colour the two halves split apart with an audible crack, wrapping backwards as they catapult the seeds away from the parent plant. Most fall within a metre, but a big pod rupturing in the heat of summer might eject seeds to a distance of 4.5 m! This distinctive and conspicuous European native grows well on Lüneburg Heide, and indeed right across western and central Europe. Broom loves the wide-open, full sun and dry, well-drained sandy environments of Europe's deserts, can flourish in exceptionally acidic soils, and can survive extended periods of drought. A leguminous plant, nitrogen-fixing bacteria (*Rhizobium* sp.) located in nodules on its roots help make this plant competitive on poor soils, this symbiotic relationship being so effective that 81% of nitrogen in its above-ground tissues is derived from the atmosphere. Broom leaf and stem litter return nitrogen to the surrounding soil, and heath farmers of the past sowed the plant to produce so-called bram groves which provided food and litter for sheep and goats,

A member of the pea family, Common or Scotch Broom is a large shrub found on sunny, dry and sandy sites at low altitudes, and can tolerate very acidic soil conditions. (Berit, CC BY 2.0)

Common Juniper, an evergreen conifer, has the largest geographical range of any woody plant, with a circumpolar distribution throughout the cool temperate Northern Hemisphere. (J.-H. Janßen, CC BY-SA 3.0)

perhaps explaining why another German name for the plant is goat's clover. Broom stands may become so densely overgrown that cutting with a tractor-driven flail followed by sheep grazing may be necessary to give reptiles, ground-nesting birds and botanical diversity in general a chance to flourish. Broom also provides pasture for bees, which pollinate the species, and many other insects also visit the flowers. In the past, its skinny branches were tied together to make domestic brooms – besoms, bundles of broom twigs lashed tied to stout poles – and the fibres used to make yarn and rope. Human consumption of the flowers and seeds is recorded but certainly not recommended, for they contain toxins which can produce nausea, vomiting and abdominal pain; nonetheless, flower buds are pickled and used as a substitute for capers, the green tops of the plant have been used like hops to give beer a bitter flavour, and the roasted seed is a coffee substitute.

Juniper is another personal favourite to be found on Lüneburg heath, though very different to broom. It's a rare treat to see these dense, burly, vaguely alien bush-trees scattered across the open *heide*. This aromatic native European shrub-tree of dry substrates is a dark-green evergreen conifer with small, sharp needle-like leaves and round berries. Growing from low elevations on pasture and abandoned land all the way to high elevations above the tree line, juniper was one of the first tree species to colonise after the last Ice Age. Common juniper cuts a distinctive profile alongside the other more meagre flora of heathland, its form varying from low-growing to a spreading bushy shrub or small tree. Mature plants may reach a height of 10 m, with individual plant profiles often appearing slender, columnar and pyramidal-shaped like the cypress-family members that they are. Slow-growing and with a lifespan of up to 200 years, juniper thrives on heathland with its ability to tolerate nutrient-poor soils, drought and frost, and the sparse open landscapes suit the plant's nature as a shade-intolerant species. Although not very palatable due to volatile oils in the needles, cones and wood, juniper is grazed by small and large mammals when food is short, particularly in winter. Juniper provides dense cover for nesting birds including Goldcrest, Firecrest and Fieldfare, and is the food plant for the caterpillars of several moth and butterfly species. The fruits are berry-like, fleshy and aromatic, maturing from green through blue-purple and finally to black over 18 months when they are the size of peppercorns. The berries are employed as a flavouring in a range of meat dishes including game and venison, but their best-known use is in the flavouring of gin. The name gin originated from the French name for the juniper berry, *genièvre*, which was adapted by the Dutch to *genever* and shortened by the English to gin. Once much more widespread, juniper is now rare and threatened across northern Germany as its habitat declines in quality and extent, though the plant thrives on Lüneburg, one of its few remaining population centres. Juniper is in steep decline in the UK too, due primarily to a lack of effective management and habitat loss. Specimens survive at only a few scattered sites, and the plant is thought likely to become extinct in lowland England within the next 50 years without conservation efforts. Regeneration is inhibited by overgrazing and heath burning, whether accidental or deliberately employed for habitat management purposes.

LÜNEBURG HEIDE MANAGEMENT

Lüneburg has been managed for millennia using a range of traditional methods. The heath continues to be managed today, for it is of importance at national and international scales, not least by virtue of its sheer scale. The goal of this management is to sustain its biodiversity and landscape interest, and with a long-term perspective recognising today's responsibilities towards tomorrow's generations. It is perhaps no surprise to find that the twenty-first-century approach to management on Lüneburg Heide reflects in many ways historic pastoral heath farming traditions. Such management work continues a long tradition of anthropogenic intervention on the European deserts, upon which their continued existence is dependent. Four key management methods are grazing, mowing of the vegetation, plaggen, and fire: burning.

Grazing is a key method of preserving the heaths of the nature reserve. Flocks of German Grey Heath Heidschnucke sheep graze Lüneburg 365 days a year, the sheep sometimes being termed a landscape manager species for its grazing encourages young heather shoots. Grazing on the heathland during the day, they are herded into pens at night where their otherwise scattered defecations accumulate. Only during the lambing season are the flocks not herded daily onto the open heath. Constant low-intensity grazing spurs the heath to vegetative rejuvenation, and trampling by sheep hooves maintains valuable patches of bare ground between heather plants. Here the reduction in shading ensures that a range of specialist plant species that need colonisation gaps have a chance to germinate, and mosses are less able to establish. Bare ground also provides basking, hunting, nesting and burrowing sites for numerous invertebrate, bird, reptile and amphibian species. Sections developing a dominant grass coverage may be selected for short-term treatment by intensive grazing. Mowing is another traditional management method, under which the heather is cropped to a height of about 3 cm and the cut vegetation removed. In the twenty-first century this is done by mechanical means, a tractor towing an agricultural mowing machine behind which is a trailer into which the small pressed bales of heath vegetation are ejected for subsequent transportation. Baled heather is sold to thatchers, with other commercial uses including as fodder for elephants.

Plaggen is a radical management method in which whole expanses of land are stripped of their vegetation, raw humus layers, and the top layer of mineral soil to a working depth of up to 15 cm. Discussed in Chapter Three, the plaggen method is used to maintain heather in its pioneer growth stages and restore heavily grazed expanses or areas dominated by Wavy Hair-grass and other species to heath. Specialist sod-cutting machines or regular excavators scrape away the surface layers, with the goal of producing tracts of bare sandy ground. Habitat management guidance indicates suggests that well-managed

A cleared tract of bare sandy ground on Lüneburg Heath.

A shepherd leads his flock onto heavily eroded heide land, c.1930s/1940s. (Lingen/Korte City Archives)

sites should incorporate between 1% and 10% of undisturbed and exposed bare ground. Plaggen cutting was reintroduced to the reserve in 1986, and over the first 20 years around 210 ha had been treated in this way at a cost of €4,000–7,000 per hectare. While little known elsewhere, the region of north-west Germany, the adjacent Netherlands and north-east Belgium is the worldwide centre of the distribution of these plaggen-type human-made soils and landscapes. Here, heathlands were historically an indivisible part of what was one of the most intensive agricultural systems in terms of labour input, with some farmers devoting more than half their working time to the plaggen system. Surface layers were repeatedly stripped of their thin vegetation cover, which was used as bedding in cowsheds and sheepfolds where they absorbed animal urine and excrement before being spread over arable fields to improve the soil. Extensive desertification resulted as the exposed sands began to drift in the wind, and in locations where plaggen extraction was undertaken over a long duration the resultant extensive areas of sand and shifting dunes were barely usable even for sheep farming. Only on the North European Plain did plaggen farming become widely established: the technique was mostly unknown in the UK, for instance.

Burning may not immediately seem an appropriate means for biodiversity conservation. Setting fire to the countryside will inevitably prove controversial, for incidents involving fire and flames are usually framed as crises – understandably perhaps – and particularly so when nature reserves are being fired. But so-called natural fire regimes are an established element of natural system processes in many habitat types around the world. The intentional incineration of vegetation was for millennia a traditional element of farming practice on most or all instances in the European lowlands before being discontinued as heath farming was progressively abandoned from the later nineteenth century onwards. Nevertheless, this ancient practice has been reintroduced for nature conservation purposes over the last few decades, notably on Lüneburg heath among other locations.

Although it may seem an extreme measure, burning is perhaps the most effective of various options for mitigating decline in the condition of heathland for conservation purposes. To appreciate the necessity for firing the landscape, it is important to recognise that the heathland environment is a plagioclimax, a biotic community whose ecological succession has been arrested and prevented from developing further through direct human intervention. Most or all the lowland heaths of western Europe were created by forest clearance in prehistoric times, which was typically achieved by methods including intentional firing. The open expanses newly made for agriculture and settlement were subsequently maintained through repeated episodes of burning as well as other means including livestock grazing. In these ways, the natural transition of the ecosystem from heath to scrub and ultimately closed woodland was arrested to provide a continuous supply of heather and grass forage for domestic herbivores. Human intervention such as extensive and regular burning maintained a distinctive plagioclimax environment the ecological succession of which was arrested, resulting in open landscapes and a distinctive range of flora and fauna quite unlike the surrounding lands.

Maintaining stands of heather plants as part of a mosaic of vegetation types is as critical for heathland conservation today as it was for providing forage for graziers in the past. Heather is typically the dominant plant of heath vegetation and an important food plant for several grazier species, providing a supply of palatable and nutritious forage on poor soils. The density and cover of heather strongly influence the environment in which all the other plants of the community live. Managing for variation in the physical structure of this dwarf shrub vegetation is an important objective for conservation purposes, because a distinct combination of heathland plant and animal species is associated with each of the growth phases of the heather plant. As they grow, the development of heather plants proceeds through a series of four growth phases, termed pioneer, building, mature, and degenerate (or 'over-aged'). Conservation management through manipulation of the life cycle of the heather plant aims to produce a diverse range of micro-habitats to ensure that the full range of heathland plant and animal species may flourish. The objective on individual sites is typically to maintain the overall heath plagioclimax habitat with discrete stands of heather plants at each of the four major growth phases as well as bare sandy ground to produce variation in the physical structure of the habitat. After several years of growth, the development of the heather plant has reached the degenerate growth phase and regeneration is required, in terms of both individual plants and the age range of each stand.

Burning is the easiest, cheapest, quickest and perhaps most natural means for the creation and maintenance of heathland. The controlled burning of aged heather is in fact beneficial in these environments, however counterintuitive it may seem. The shoots of young plants have the highest concentrations of nutrients, the levels declining with increasing age after about seven years of growth as a greater proportion of their biomass is in the form of woody stems and branches of low nutritional value. The objective of heathland burning is to induce the heather plants to regenerate by incinerating their above-ground biomass while leaving the roots intact to sprout once again. The historic purpose of this method of preserving the plants was to revitalise the quantity and quality of heather to provide grazing livestock with a continuing supply of forage while also countering the invasion of undesirable woody and other species. Crucially, this traditional method of preserving the plants for domestic graziers aligns very effectively with the goals of enhancing biodiversity. Authorities recommend that 5–6% of a heathland should be fired annually. Common heather is fire-tolerant, and the flames sweep over the plants in a few seconds. Almost all the above-ground parts including woody stems and branches are destroyed, but the rootstock and soil seed bank will usually withstand the conflagration at the soil surface if conditions are sufficiently moist. Naturally adapted to recovering after fire, pioneer heather soon reappears over the burnt area and the plants will regenerate, resulting in an even-aged stand.

Other elements of heathland flora and fauna are also adapted to fire in evolutionary terms. For instance, the germination of broom and gorse is stimulated by fire, and the decline of the Heath Grasshopper on heathlands in central Europe is thought to be related to the cessation of regular burning leading to the loss of its preferred habitat of sparse vegetation dominated by grasses and patches of bare ground under trees, leaf litter and moss. Regular firing prevents the build-up of fertility in the ecosystem and reduces the volume of combustible matter on heathland sites, mitigating the severity of accidental fires. However, burning may be unsuitable ecologically for certain sites, potentially leading to a net loss of biodiversity notably including reptiles and amphibians. Small sites and those in close proximity to woodland or houses and urban development may be unsuitable for burning, and in addition it may need to be carefully planned to preserve the archaeological record.

The application of prescribed burning in combination with other management methods to maintain heath and grassland was for long traditional in Germany. Today, burning on Lüneburg Heide is planned in places where the raw humus layer is not overly deep, and executed typically in the months January to March either on blocks of land a few hectares in extent or in the form of linear corridors snaking across sites. Ignited with a mixture of petrol and diesel from a hand-held drip torch, the flames propagate across the ground, the spread of the fire held in check by means of firebreaks mown around each block or corridor and perhaps too a tractor-towed bowser spraying water onto the heath a few metres upwind of the fire-starters. Once the fire is over, the blackened expanses are ready for fresh growth. Heather typically re-establishes within one to five years following a burn, varying with the state of the vegetation pre-burn and soil moisture. Unsightly charred expanses of bare ground will normally disappear under a closed canopy over that time, though if a 'hot burn' destroys the upper humic surface including the rootstock and soil seed bank there may be a need to introduce a seed source. In practice, given the scale of Lüneburg the single most significant element of habitat management is probably the several thousand Heidschnucke sheep and goats grazing there. Nonetheless, grazing alone is seldom adequate to prevent succession towards woodland and is considered effective over the long term only when implemented at a high intensity, a tactic which however may conflict with other ecological goals on individual sections.

The rationale for burning heathland has been strengthened in the twenty-first century as atmospheric nitrogen deposition has resulted in significant eutrophication – nutrient enrichment – of the soils. In removing 50–80% of the biomass, combustion eliminates large quantities of nitrogen from the system, with about 90 kg of nitrogen per hectare directly discharged, depleting nutrients which may have otherwise supported less-favoured successional species including Wavy Hair-grass and other invasive plants. Further nitrogen discharge occurs in the year after the fire as a result of mineralisation and leaching on the burned area. In contrast, these undesirable plant nutrients remain on the ground surface after grazing and mowing treatments.

ECOLOGICAL MANAGEMENT AND PUBLIC OPINION

Deliberately setting fire to the countryside frequently proves controversial. People's attitudes towards prescribed burns carried out on cherished local nature reserves are understandably often negative. Communities are often unable to reconcile this seemingly wantonly destructive management method with the goal of conserving biodiversity and enhancing natural beauty, and the risk of an uncontrolled conflagration inevitably colours peoples' opinions towards this means of ecological management. Many are unaware that the origins and characteristics of many treasured landscapes are semi-natural, having been formed and maintained by human agency interacting with natural processes. Our largely urbanised society has long forgotten the traditional role of fire and other ecological management methods in our landscapes. Places intentionally maintained as open treeless wastes may be regarded as unattractive in terms of landscapes and as environments for recreation – and particularly when firing is involved, perhaps bringing to mind uncomfortable parallels with instances of large-scale deforestation around the world. Grazing and mowing may be more acceptable to local residents, though even this is by no means certain. The debate centres around meanings and perceptions of nature, naturalness, and nature conservation, with stakeholders' thoughts typically focused upon familiar vistas while lacking knowledge about the need to protect nature in terms of species, many of which are often unspectacular, or even barely visible. Arresting or retarding ecological succession to maintain a plagioclimax by means of fire may be viewed as intolerable interference with nature, or God's intentions for the world. Conservation professionals may be perceived unfavourably, viewed as outsiders from the 'conservation industry' dominating debate. Their values, perceptions and professional opinions may be interpreted as overriding those of local residents; their proposed goals and objectives for management regarded as externally derived 'desecration' and 'eco-vandalism' imposed upon sites and communities; and individuals denigrated in personal terms.

So as heath farming finally ceased as an economic activity, the formerly widespread cultural practice of burning the lowland wildernesses of Europe dwindled too. After a history probably stretching back several millennia, by the later nineteenth century heath burning had become an archaic practice associated with earlier, perhaps more primitive times. Nonetheless, for these plagioclimax habitats and their biodiversity to have a future, anthropogenic intervention is a necessity, and prescribed burning for conservation management continues today in some locations. Given the merits of the method, the once-widespread European tradition of prescribed burning may need to be re-established and practised more widely despite its controversial nature, in combination with various permutations of grazing, mowing and plaggen.

Here, the path continues south from Lüneburg Heide, opening fresh heathland vistas. No wonder the Heidschnuckenweg was nominated a decade ago as the most beautiful trail in Germany. These German deserts are extraordinary for their generally favourable ecological condition and the sheer extent of these wide-open expanses with their palpably raw and elemental atmosphere. When the sun shone, the Lüneburg environment warmed rapidly and I was again reminded just how little shade these heathlands offer. You wouldn't want to be here in stormy weather, either. After several days hiking, which were rarely anything less than a treat, I quit the 'Way of the Sheep' and headed for a nearby rural *Bahnhof*, where I encountered a passenger train service the rolling stock of which was emblazoned with the brand 'Der Heidesprinter': so, a train paying tribute to the local heath. Feeling triumphant and charmed, and having developed a real love for these rolling German wilds, I realised I'd learned far more about the deserts of Europe than I might ever have reasonably expected.

For long the German *heide* lands dominated much of the North European Plain. It is sobering to appreciate that, for many peasants over many centuries, these were the only landscapes familiar to them: they were born on heathland, lived and died in these environments, and very likely experienced few or no others. Today it is instructive to observe just how capably the Germans are conserving their extensive remaining desertic wastes for biodiversity, landscape and other goals. The once-deprecated and scruffy though glorious purple heather-clad *heide* wildernesses enjoy popular recognition and respect as treasured components of German national heritage, and a broad spectrum of conservation measures have gained public support. Now a significant nature tourism destination, Lüneburg Heide is promoted with contemporary branding methods, carefully planned visitor management minimises conflict with conservation priorities, and links have been made with local hospitality, food and other businesses. In times when most people's daily landscapes are becoming increasingly similar visually, these outlandish desertic vistas inculcate a distinct sense of regional identity. Perhaps the German approach offers a model to be adopted for the many other threatened arid areas of Europe in the twenty-first century.

Conclusion

EUROPEAN DESERTS: PERSPECTIVES, AND A VISION OF THE FUTURE

Europe has deserts. Unexpected, perhaps, but true. This work has identified and explored several examples of the numerous such arid territories scattered across the subcontinent. A century and more ago there were many more of these lowland wildernesses; those we see today are the last remnants, and though valuable for a host of reasons they are mostly unknown and their continuing existence is threatened. A central element of the intellectual contribution of this work has been to conceptualise these mostly barren tracts in aggregate as deserts located in inland Europe. It advances the proposition that there is an arid Europe set amidst the familiar verdant one; that it is of inestimable value; and that it must be conserved collectively and on its own merits.

Earth is the blue planet. Most of the Earth's surface comprises water, yet it is easy to overlook the fact that terrestrially the liquid is in short supply. Dry land is ubiquitous over the face of the planet, most notably of course in its deserts – the Cinderellas of the world's environments, these frequently obscure, neglected, marginal wildernesses. A large proportion of the planet's deserts are located in Eurasia, and at its western extremity verdant Europe has its share. Though until now barely acknowledged, these desert tracts where vegetation is scarce or absent are part of the traditional landscape of lowland Europe. Europe's cold deserts are distinguished with reference to soil aridity – edaphic aridity – rather than primarily high ambient temperatures. Aridity manifests in cold climates as readily as in hot, and water scarcity is a key attribute of these territories, alongside their predominantly mineral soils – nutrient-poor, infertile and acidic, and frequently set on sand and gravel. Formerly much more commonplace across the subcontinent, the number of sites remaining is unknown but is thought to be in the order of thousands. While their total present-day extent is estimated to be less than 1% of terrestrial Europe – so much has been lost – their contribution to biodiversity, landscape and the cultural history of the subcontinent is of far greater significance than may be inferred from their aggregate area.

Collectively, their distinctive features distinguish the European deserts as environments which differ fundamentally from their verdant surroundings. While mostly set under a primarily temperate maritime climate regime they are often the locus of unfamiliar environmental extremes including high temperatures, soils deficient in moisture and with intense solar radiative regimes at the surface, drifting sand and active dunes in some locations, and sandstorms, among others. Their landscapes vary but are mostly unenclosed and open in aspect, typically exhibiting lowland flats and plains, low hills, and sometimes flat-topped plateaus and mesas. Historically characterised as wastelands, they are mostly agriculturally marginal and have few or no other conventional economic resources to offer. Barely and even unvegetated in some instances, the European deserts support a biodiversity which, while relatively depauperate, is unique, with many rare and endangered species unable to survive elsewhere. Many have been designated as nature reserves, nature parks, national parks and similar – though many remain undesignated – but nonetheless they are in general vulnerable, frequently neglected, contracting in extent and declining in biodiversity value. Typically remote and mostly little known and under-appreciated, Europe's arid territories are typically barely or unpopulated and frequently share common histories as low-status localities associated with social marginalisation, poverty, lawlessness and risk. Finally, while few or none are formally named as deserts, the term has regularly been informally applied to these dry, sparse territories.

The seven country chapters illustrate that, while the deserts of Europe evince numerous commonalities, they are also diverse. Ultimately, every site is unique. Bardenas Reales in Spain offers a spectacle of desertic badlands like few other locations in Europe; the parched Dutch yellow-sand wildernesses are striking in their stark weirdness; in France, the Landes desert – which constitutes an entire region – may now be mostly afforested but nonetheless the wild sands beneath are still palpable; the 'Polish Sahara' at Błędów is a vast unvegetated tract of sand with high temperatures, sandstorms and occasional mirages; England's desiccated Breckland is a scarce haven for rare species in a region of intensive farming; and Lüneburg Heide is a superb survival of the once-extensive rolling heathery landscapes of Germany. Finally, Iceland, an outlier in almost every sense, trumps them all. Nowhere else in Europe do deserts have a similar recent volcanic genesis, or present sandy wildernesses of such otherworldly barren intensity and on an unparalleled scale. Given the exceptional attributes of Iceland's arid tracts, many of the syntheses, analyses, conclusions and recommendations presented in subsequent sections here are not relevant to the largest desert in Europe.

(As author, I feel bound to acknowledge here the limitations of my knowledge and understanding represented by the innumerable European deserts I have yet to explore. And, while on the subject, one's imagination inevitably drifts to wondering about the vast former expanses that have been lost over the millennia without trace or record. What enchanting vistas the European deserts must have offered in their pomp.)

The European deserts are valuable for what they are and what they represent. A relict manifestation of lowland wilderness, the European deserts bring incongruous desertic vistas to otherwise verdant landscapes. Collectively they constitute one of Europe's native range of environments and habitats, though one hitherto only infrequently acknowledged as such. There is nowhere on the subcontinent quite like these places, and Europe's environment is all the more interesting for their continued existence. Representing the last vestiges of the subcontinent's formerly extensive wastes, and some of its least agriculturally favourable land, they are valuable collectively as the subcontinent's representatives of the vast arid deserts strewn across the planet. The western worldview has for centuries viewed the conquest and subjection of nature as a major challenge. Over the course of several millennia, these and other similar marginal lands have progressively been transformed to support conventional agriculture, to the extent that almost everywhere that could have been improved with modern agricultural methods has been; yet still agriculture is identified by the European Environment Agency as continuing to severely impact these habitats. Nonetheless the remaining deserts endure, reaffirming the fundamental proposition of the dominance of the planet's physical environment over the human. The preservation of species, populations, ecosystems, biodiversity and nature more generally should be of human concern, and these places serve as increasingly scarce refugia for numerous specialist species adapted to arid conditions with few or no alternative options for existence. The deserts of Europe make a critical and irreplaceable contribution towards the goals of maintaining the biological diversity of the subcontinent and mitigating the risk of species extinction. They are also significant collectively as a physical expression of a common European heritage – both environmental and cultural – though their histories have seldom been studied in any detail. Paradoxically perhaps, military use has sustained the existence of substantial tracts, places where conservation goals may better be achieved given the level of resources typically available for their management, though these are rarely open to the public.

Conclusions drawn in the few available assessments of the conservation status of Europe's surviving deserts are unfavourable. The threat is of a significant reduction or loss of biological diversity, and the prospects for the future seem poor. Their collective conservation status has been assessed by the European Environment Agency as predominantly unfavourable across most of the subcontinent, with a majority of instances in the countries examined here deemed in poor or bad condition. Relatively few sites were assessed to be of good conservation status, notably some of those located in Germany and the Netherlands. The conclusion drawn by the European Environment Agency is that the overall conservation status of the European dry heath habitat – employed as a surrogate for the range of habitats collectively treated as the European deserts for this work – is unfavourable and not improving.[1] When the target land classification is broadened to the 'Heath and Scrub' habitat group, only about one-quarter of sites are assessed as having favourable conservation status, with about half classified as 'unfavourable-stable' or 'unfavourable-deteriorating'.[2] If they are lost, nothing can replace them.

---

1   Olmeda, C., Šefferová, V., Underwood, E., Millan, L., Gil, T. and Naumann, S. (eds) (2020) *EU Action Plan to Maintain and Restore to Favourable Conservation Status the Habitat Type 4030 European Dry Heaths.* European Commission, Directorate-General Environment.
2   European Environment Agency (2020) *State of Nature in the EU: Results from reporting under the nature directives 2013–2018.* EEA Report No 10/2020. Luxembourg: Publications Office of the European Union.

EUROPEAN DESERTS: PLANNING AND MANAGEMENT FOR THE FUTURE

Europe's deserts require planning and management to ensure long-term survival of the habitat and its characteristic flora and fauna. That they have survived at all is because they have been managed. To repeat, many or most are semi-natural rather than wholly natural in nature, and they are dynamic, meaning that maintaining a state of equilibrium necessitates continual management input, as is the case with other habitats. Many sites are legally protected as nature reserves – though a good deal are not – and mere designation frequently does little or nothing to ensure adequate management. A full discussion of planning and management is beyond the scope of this work, and management prescriptions must reflect natural and regional variability. However, nutrient stripping in its various forms has for millennia been a foundation of the land management regime which historically sustained the deserts, constituting another element distinguishing them from the conventional agriculture of their surroundings. A critical management issue centres upon how to maintain and restore the low soil nutrient levels of Europe's deserts, which should continue to be a key objective of management and one requiring active and continual effort along with a sustained allocation of resources over the long term.

Some commentators consider that, given their semi-natural nature and origins, the only reliable option to maintain the European deserts is to farm them systematically once again. The view is that, without restoration of their former agricultural production functions they will inevitably decline in extent and quality over the long term; and that their long-term conservation may best be achieved through reinstating traditional practices, acknowledging their nature as primarily cultural landscapes. Extensive grazing was one of the main traditional uses of these tracts, and is particularly suited to the management of large areas today because it reduces the need for costly mechanical or manual means of scrub control. The notion of reintroducing the farmer as heathland manager clearly has its merits, though this role must encompass a broader range of objectives including promoting biodiversity and enhancing the landscape in addition to the sole conventional goal of food production. However, farming on the European deserts became progressively less economic many decades ago, little or nothing significant can be done to modernise sheep-keeping, and subsistence agriculture with its hard work, long hours and low wages is not a viable economic model in the twenty-first century. Few of the available habitat management options produce significant returns, meaning – again – that a suitable subsidy regime is required, one potentially entailing an indefinite commitment of resources.

Reversing the historic trend of heathland loss through habitat re-creation offers a further strategy for their long-term conservation. The goal is an ambitious and radical one: simply, to produce more, bigger and better-connected spaces for wildlife to flourish. Heathland re-creation involving large scale conversion of land from other uses is presently in progress in several European locations. The initial focus of habitat re-creation projects is usually upon areas known to have supported heath vegetation in the past, typically land adjacent to existing tracts and presently down either to crops, improved grassland or plantation trees. The nature of the antecedent land use is an important determinant of the likely success of habitat re-creation. On sites presently under arable crops, and to a lesser extent improved grassland, efforts to recreate lowland heath may achieve only limited success in the short term at least, for agricultural use will not only have reduced or eliminated heathland habitat species but will usually also have caused significant changes to soil properties. Treatments applied to agricultural land in past times in the form of lime and fertiliser can cause long-lasting changes to soil chemical characteristics which combine with changes in soil structure to make substrates unsuitable for the growth of heathland species. Increased soil fertility tends to favour an abundance of undesirable rapidly growing species which are effective competitors during the early stages of heath re-establishment. Also, the progressive decline of the heathland seed bank buried under such sites may present a further constraint. Site treatment may be necessary before attempts are made to reintroduce heathland species, employing methods including continuous cropping to deplete soil nutrients, turf or topsoil stripping, and the addition of pH-modifying amendments.

Land presently under tree plantations offers significantly better prospects for habitat re-creation compared to arable sites. Here the conditions necessary for heathland re-creation are more likely to be met, notably with reference to soil nutrient levels. Removing conifers and other trees offers the most practical and cost-effective means of re-creating heathland expanses in the short term, and the quality of habitat recovery achieved on former plantation sites in the UK and elsewhere has been impressive. The physical and chemical properties of soils lying beneath both planted and self-sown conifers are similar to those of undisturbed heathland, for the impacts of plantation forestry upon soil properties and the buried heathland seed bank are generally much less detrimental to the prospects for rewilding. Viable heath seed banks frequently persist under both conifer and broadleaf woodland, in some instances for periods of 40 years or more. Intervention to manage invasive tree regrowth is frequently necessary after felling operations, and removal of excess litter and brash may assist the regeneration of heathland species. Ironically, the large-scale conifer planting policies implemented a century or more ago by many European governments have perhaps offered an unlikely lifeline to some of the subcontinent's lost desertic tracts, for without them the fate of many lowland heaths may have been conversion to agricultural use in the mid-twentieth century with the result that the stock of suitable land potentially available in the twenty-first for nature recovery would be much reduced.

This work proposes the target of a doubling of the total area of the European deserts by the middle of the twenty-first century to serve habitat and species recovery goals. This would clearly be a major endeavour, and one which would need to be planned at the subcontinental level. Ecosystem re-creation on such an extensive scale is merited given the significance of the habitat and the catastrophic losses it has sustained over the past century or more. Public policy towards the land-based industries has historically taken priority over that devoted to nature conservation – at least in the UK – and the prolonged period of public investment in the intensification of agriculture and forestry incentivised the loss of heathland across much of lowland UK, but times are changing. A doubling of the extent of heath habitat across the subcontinent by 2050 is clearly an ambitious target, but the needs of biodiversity on the European deserts are most effectively addressed on a large scale. The goal is in accordance with the historic global agreement to halt and reverse nature losses reached at the 2022 UN Biodiversity Conference (the Kunming-Montreal Global Biodiversity Framework). This agreement includes a target – colloquially known as '30x30' – for one-third of the earth's land and sea to be conserved by 2030 through the establishment of protected areas which are to be effectively conserved and managed, and connected by ecological corridors. The process of assisting the recovery of an ecosystem that has been degraded or destroyed is termed ecological restoration. Described as the biggest conservation commitment the world has ever seen, the 30x30 target is clearly of relevance to Europe's sandy lands, an opportunity to reverse the historic trend of heathland loss through carefully planned and managed re-creation to create large areas of new habitat – more, bigger and better-connected spaces for wildlife. The goal of doubling the area of desertic lowland wilderness across the subcontinent by 2050 should be pursued in parallel with the improvement of the condition of those presently in existence. Rewilding projects may conceivably bring nature conservation into conflict with national agricultural and forestry priorities, although the loss of output as land adjacent to existing tracts is taken out of production is likely to be negligible. It is ironic that re-creation of these 'waste' tracts is now on the policy agenda across the subcontinent, though mostly unquantified. The time is now.

Public knowledge of and attitudes towards Europe's deserts may hinder their successful conservation. Many people are unaware of their existence, and many do not appreciate their significance as valuable habitats. In the lowlands, meadow and woodland are perhaps typically regarded as attractive norms for landscapes deemed natural, while heath is frequently regarded as significantly less attractive, judged to be unkempt, derelict-looking and even hazardous. Habitat management work on woodland sites to restore former desertic tracts frequently draws criticism, involving as it does intense and seemingly destructive activity as trees are felled, leaving desolate newly cleared vistas in its wake. Notions of God, naturalness and divine creation hold that any human intervention in the natural world is undesirable and wrong-headed. Local communities and recreational visitors may consider any landscape change in the countryside undesirable and unsettling, particularly if it involves unfamiliar heavy machinery and is perceived to have been externally imposed by agents including conservation professionals (whose hard work routinely goes unappreciated, for the wildlife – the main beneficiary – never says 'thank you' either). Consequently, there is a need to improve the public's awareness and comprehension of Europe's deserts to support the argument for their preservation; and also to gain recognition in broader terms of the fact that all habitats require appropriate management.

In cultural terms, Europe's deserts are as worthy of conservation and celebration as are much grander and more conventional human-made cultural artefacts like the subcontinent's numerous palaces, cathedrals and artworks. Both require the investment of adequate resources for their maintenance, without which they will deteriorate. The fact that the European deserts are predominantly cultural landscapes in no way diminishes their importance or significance in scientific terms. More than three-quarters of the terrestrial biosphere has been reshaped into anthropogenic biomes – anthromes

– through humans' use of land, including the European deserts. Valuing anthropogenic landscapes like these is as essential for biodiversity conservation as is preservation of the world's remaining wild areas. Maintaining such traditionally managed landscapes, which harbour a rich native biodiversity, is important to prevent losses from a species pool which may stretch back to prehistoric times. In addition, since evolutionary adaptation within species to local environments is widespread in the natural world, the European deserts may also constitute important reservoirs of the genetic resources of species adapted to survival under arid environmental conditions in a world experiencing climate change.

The social history of the European deserts is not well known. The written record is scant, much needs to be researched, and the voices of those formerly living on or from these frequently poverty-stricken tracts seem particularly scarce. One dominant sociocultural theme consistently associated with them is that, historically, they were commonly regarded as low-status localities. With the worst and least valuable land, their inhabitants were impoverished in every sense. While these territories were inhabited and farmed for thousands of years, and their land-based products played a significant role in at least some regional economies, both their landscapes and peoples were commonly despised by respectable society, with their inhabitants regarded as uncivilised at best and inherently criminal at worst. State-driven efforts to tame and monetise Europe's desertic wilds have historically included land settlement colony projects and work camps operating under managed regimes of a type mostly unknown in other locations. These have been conceptualised by some as manifestations of a political philosophy of internal colonisation, 'planting' new populations on the wastes in a manner comparable to the creation of overseas colonies under historic imperial administrations. One of the implications of their existence is that the traditional rural inhabitants had been deemed unwilling or incapable of managing their local environmental resources in a suitably productive manner, and that repopulation with newcomers was necessary. Ironically, today many of these once-shunned localities have become desirable and even gentrified residential locations.

A NEW EUROPEAN DESERT?

It was a big surprise to discover a new European desert in the process of formation. Southern Romania is turning into a desert. In south-east Europe, some of the drivers that were instrumental in the historical formation of European deserts continue to prevail in the twenty-first century. Sited upon fine aeolian sand, the wide-open plains of Oltenia in the country's most arid region are hot and dusty. Here, sand has been advancing over hundreds of hectares of fertile land each year, blown by the wind onto crops, roads and into people's houses. Two-thirds of the region's population works in agriculture, but grapes frequently fail to ripen on these parched sands with little capacity for water retention. In the 1960s a large-scale national project to boost Romanian agricultural production saw thousands of hectares of forest felled here and the formerly extensive sand dunes levelled and cultivated. Drought-resistant acacia trees were planted as wind breaks to impede the desiccating effects of the wind, and a new irrigation system introduced. Agriculture flourished, but the soil was difficult to cultivate and the result of removing the forests, which had served to diminish the regional effects of aridity, was lower humidity and more desertification. When the state returned land to the people after the 1989 Romanian Christmas Revolution, 30,000 ha of forest including acacias were thoughtlessly felled and the irrigation system fell into disuse.

Here at the westernmost extreme of the Eurasian steppe desertification has accelerated wind erosion and now impacts an area of 100,000 ha along a 30 km-wide strip of territory near the border with Bulgaria. Sandstorms are frequent, sometimes reaching Bucharest 200 km to the east. The arid land around the small town of Dăbuleni is increasingly barren, and in response some farmers have turned to non-traditional crops including sweet potato, peanut, date, olive and almond on land that formerly produced orchard fruit, grapes and cereals. As arable productivity declines and some areas have become unusable for crop production, land is abandoned and the people of the 'Oltenian Sahara' outmigrate. Against a background of climate change, regional temperatures are projected to rise by up to four or five degrees over the twenty-first century, bringing high soil temperatures and oppressive heatwaves, and making soil water deficits and summer drought increasingly common. Here is an extraordinary phenomenon: a modern-day desert in the making in Europe, similar to those forming in other locations around the world. Over the next 50 years, much of southern Romania may be covered with sand. Shrewdly, an informal open-air sand museum was founded in Dăbuleni to preserve a 12 ha tract of the powder-white sands which was spared from cultivation; around the town lie more than 3,000 unproductive hectares. Volunteers have planted thousands of acacias on the sandy terrain and reforestation is being considered, but it is unclear whether the region's dramatic and relatively rapid environmental transformation to a human-made desert can be successfully halted. With about 70% of Romania's land vulnerable to desertification linked to climate change, it seems clear that the issue is not restricted simply to the subtropical regions of the planet. Desertification is a contemporary reality in Europe, too.

EUROPEAN DESERTS AND THE ANTHROPOCENE

How do the genesis and continued existence of the European deserts relate to the contested Anthropocene thesis? A descriptor of human impact on the Earth system, notions of Anthropocene are founded upon the assertion that, over time, human activity has overtaken natural processes to become the dominant influence upon the planet. Humans are situated as agents responsible for manifesting transformative change in earth systems, and their activities can be traced through virtually all planetary processes. A substantial proportion of the Earth's terrain has been shaped by human activity throughout the course of the Holocene and particularly during the modern period. Many or most of today's European deserts date from the period following the last Ice Age as *Homo sapiens* cleared woodland on light soils for farming, fuel and materials. Prehistoric populations had a significant and widespread impact on the environment of the subcontinent over several millennia as the fertility of the fragile soils was exhausted. Whole tracts were transformed, the influence of human activities varying over time in response to population pressures, local and national politics, changing technologies and climate change, among other factors. Driving change from one biotic community to another resulted in profound changes in soils and landscapes together with a fundamental transition in the mix of flora and fauna species. Human intervention became the dominant influence upon these tracts in the sense that the Anthropocene thesis proposes, wreaking transformative change in earth systems which entrenched the present-day biodiversity-impoverished state of the European deserts. Extraordinarily, perhaps, the consequences are still tangible today.

Humans changed the Earth. The European deserts may be considered a product of the sort of anthropogenic-driven impacts upon natural processes that the Anthropocene thesis posits. The magnitude of the ensuing impacts – in this instance the genesis and continued existence of the European deserts, these predominantly cultural phenomena – offers support for the general proposition under the Anthropocene thesis that human activity may indeed become a significant influence on the environment. Collectively, the heaths may represent some of the early phases of the Neolithic agricultural transition, the long process of anthropogenically driven land cover change which has continued to the present day. Their history implies that the Anthropocene commenced many thousands of years in the past, far earlier than the 1950 date often cited. It is instructive to recognise that only relatively primitive technologies were sufficient to produce large-scale and sustained environmental impacts across the subcontinent.

EUROPEAN DESERTS AND CLIMATE CHANGE

What impacts may climate change have on Europe's arid habitats? Climate affects most areas of life, directly or indirectly, and it is changing. Biodiversity is already responding to climate change, and the resultant environmental changes are disrupting natural habitats and species in numerous ways. Climate change is acknowledged as one of the drivers of change in these arid habitats, and its impacts are expected to intensify over time. At the global scale, most of the world's land has become drier in recent decades. Increasing aridity is influencing almost every region of the globe. Global dry lands expanded by about 4.3 million square kilometres over the three decades leading up to 2020, an area equal to half the size of Australia. The drying tendency has been particularly prevalent in Europe, and during the twenty-first century 15–20% of the land is expected to become significantly more arid, particularly in the Mediterranean region where large areas are facing the risk of desertification. Past periods of sand mobilisation and drifting on the Europe deserts have been associated with rising temperatures during historic episodes of climate change, raising the possibility of similar consequences in future as the global climate tends to greater aridity.

To focus upon the UK, all areas are projected to become warmer over the twenty-first century, with a change towards warmer, wetter winters and hotter, drier summers and with the greatest change in projected summer temperatures occurring in southern England. Rainfall totals may change only little overall, but the annual balance will shift with more in winter and less in summer, and a greater proportion will fall in the form of heavy downpours. Dry lowland heath is considered to be highly sensitive to climate change, which represents an additional threat to the habitat's long-term viability. The implications for these environments include:

- temperature: higher temperatures and more frequent droughts will intensify heat stress and may make arid soils more vulnerable to wind blow
- higher temperatures may change competitive relationships, facilitating the dominance of grasses, bracken and other species as they outcompete nutrient-poor substrate specialist species, potentially leading to a habitat shift from heathland to acid grassland with associated species loss
- fire: a key risk, wildfires are more likely in a climate with higher temperatures and more frequent droughts
- soil moisture: warmer summers with more frequent and intense droughts will accelerate evapotranspiration, leading to dwindling soil moisture levels in a habitat sensitive to changes in hydrological conditions. Native species of wet heathland such as Cross-leaved Heath are threatened, and even common heather plants may not survive periods of severe drought
- eutrophication: wetter winters may accelerate the atmospheric deposition of nitrogen, a major driver of change from a low nutrient and infertile habitat to grassland
- soil erosion: caused by heavy rainfall during major storm events.

Heathland acts as a carbon sink, sequestering and storing more carbon than modern agricultural landscapes though typically less than peatlands, saltmarsh and established woodlands. By actively sequestering carbon dioxide gas from the atmosphere, the habitat has a role to play in climate change adaptation. In this and other ways, Europe's deserts serve as natural capital – nature providing benefits to people – further contributing to the rationale for their conservation.

One interesting dimension of climate change is its potential impact upon the distribution of organisms as climatic zones shift across Europe. British heathland species presently at the northern extremes of their geographical distribution may benefit from an expansion of their range as the country's climate becomes milder, including Smooth Snake and Sand Lizard. Heathland species presently distributed in southern Europe are likely to extend their ranges northwards, including for example the Dartford Warbler. Historically distributed in southern Europe and only infrequently sighted in the UK, the numbers of this small brown bird species have increased across northern Europe over the past half century in a phenomenon attributed to the trend of increasingly mild winter conditions. The bird's core UK range has expanded from the southern counties of Dorset and Hampshire to more northerly areas including Suffolk and Derbyshire. Climate-based projections for the Dartford Warbler indicate that by 2080 more than 60% of its current European range may no longer be suitable for the bird, and severe population declines in Spain and France suggest that this is happening already. The European deserts may conceivably become refugia for a range of flora and fauna presently distributed in the Mediterranean and North African regions as climatic zones shift. The implication for the planning of conservation on these sites is that it should incorporate the needs of climate change adaptation. When determining conservation priorities, it will be vital to integrate those that address the broader international dimension alongside priorities focused upon the local and regional context.

EUROPEAN DESERTS: PERSPECTIVES, AND A VISION OF THE FUTURE

Westruper Heide, western Germany. (Evgeni Tcherkasski)

The lowland deserts we see today are the last fragments of a type of landscape which was once a familiar sight across the subcontinent. An integral though unlauded component of Europe's environmental heritage, these fascinating desertic tracts should be better recognised and acknowledged for what they are. This work has endeavoured to conceptualise and introduce them, identify and explore a handful, and enthuse the reader to investigate and cherish their charm and magic. Scruffy and elemental, there is something liminal about these incongruously barren bucolic backwaters set amidst a verdant subcontinent, bringing to Europe a frisson of the exotic, the romantic, the extreme, and a glimpse of the planet's vast arid desert wastes – though one disconcertingly close to home. It must be acknowledged that humans did not create the European deserts, but we have exerted a profound influence on their persistence and the character of their ecosystems. They offer a tantalising glimpse of landscapes and land uses predating modern agriculture, and exemplify how disparate European societies developed ingenious and broadly similar approaches to extracting a livelihood from some of the subcontinent's most unpropitious lands. Their unique biodiversity is a byproduct of physical conditions and several millennia of pastoral farming, the human agrarian tradition which provided the continuity for their ecological processes, and as cultural landscapes they are valuable as an element of the common cultural heritage of the subcontinent. Today these dynamic habitats endure as living, tangible and distinctive semi-natural environmental realms presenting magnificent wide-open vistas of tranquil native wilderness. The ambience of the European deserts was described positively by Raunkiaer:

> The warm breeze during its leisurely passage over moor and heath has become imbued with their delicious aroma. Gently and steadily it sweeps across the immense expenses, which offer it no resistance. Peace and quiet reign supreme.[3]

With timely, planned, concerted and sustained action, these exceptional environments will continue to survive and thrive into the future for the sustenance of humans and wildlife alike.

---

3   Raunkiaer, S. (1934) *The Life Forms of Plants and Statistical Plant Geography*. Oxford: Clarendon Press.

VISION FOR THE EUROPEAN DESERTS

This final section presents the author's proposed vision for the European deserts in the year 2050. By halfway through the twenty-first century, they will be flourishing as an integral and cherished component of the subcontinents range of habitats, at a favourable conservation status and with their aggregate area doubled.

By the year 2050:

- The notions that much of the planet is arid, and that deserts are distributed from the equators to the poles, have achieved broad recognition.
- The deserts of Europe have collectively achieved widespread recognition and acceptance as a valid, integral and valuable component of the range of native terrestrial ecosystems across the subcontinent and wider planet.
- A successful outreach campaign has raised public awareness, informing and educating people about the habitat's origins, history and significance for biodiversity and society in the context of European and global biodiversity decline.
- A comprehensive pan-European vision statement has been produced for the deserts, linked to a strategic plan for their conservation at the biogeographical level which reflects the variability of these tracts across the subcontinent.
- The goal of doubling their combined extent across the subcontinent has been successfully achieved.
- A management plan has been produced for each instance with clear conservation objectives – including provision for recreational use where appropriate – and is regularly updated.
- Loss of habitat and biodiversity on the European deserts has been halted; land-use changes that would negatively affect them have ceased; all tracts are legally designated, fully protected and actively managed as nature reserves.
- Sufficient resources have been secured to support long-term recurrent management interventions aimed at maintaining the ecological quality of the European deserts.
- Restoration measures have improved the ecological quality of all existing European desert tracts to favourable condition.
- Habitat recreation work has re-established large tracts which had been historically lost.
- Fragmentation has been minimised and ecological connectivity restored.
- Extensive farming systems have been re-established where appropriate in support of conservation priorities.
- Actions have been implemented to make the habitat more resilient to climate change and to fulfil its potential for carbon storage.
- Issues around degradation of the habitat through atmospheric nitrogen deposition have been addressed.
- Closer relationships have been established between the European deserts and their neighbouring human communities and local economies.

- Appropriate European deserts have been successfully marketed as ecotourism visitor destinations consistent with wildlife conservation objectives.
- A pan-Europe management forum operates to coordinate and represent the cause of European deserts conservation; acts as a forum for the exchange of experience on the effectiveness of habitat management approaches; and has produced a set of best-practice guidelines allowing for local variability.
- The history of the European deserts has been extensively researched, their landscape biographies studied, their rich social and archaeological heritage explored, and their artistic and cultural heritages documented, interpreted and published.
- Success is being actively monitored with reference to defined targets.

By the year 2050 the deserts of Europe are valued on their own merits, with their biodiversity intact and their collective future assured. They survive and thrive as a distinctive though scarce manifestation of native wilderness on the subcontinent, and collectively as Europe's representatives within the global arid desert phenomenon, reaffirming the fundamental proposition of the dominance of the planet's physical environment over the human. Europe has deserts.

BIBLIOGRAPHY

Introduction

Alonso, I., Holloway, J., Ede, J. et al. (2009) *Guidance on protecting soils and the historic environment when restoring or re-creating lowland heathland. Natural England Technical Information Note TIN054.* Accessed at: http://publications.naturalengland.org.uk/publication/33016 (3 May 2020)

Díaz, J., Linares, C., Carmona, R. et al. (2017). Saharan dust intrusions in Spain: Health impacts and associated synoptic conditions. *Environmental Research*, 156, pp. 455–467. https://doi.org/10.1016/j.envres.2017.03.047

Ezcurra, E. (ed.) (2006) *Global Deserts Outlook*. Nairobi: UNEP.

Gray, M. (2019) Geodiversity, geoheritage and geoconservation for society. *International Journal of Geoheritage and Parks*, 7(4), pp. 226–236. https://doi.org/10.1016/j.ijgeop.2019.11.001

Hunt, N. (2021) *Outlandish: Walking Europe's Unlikely Landscapes*. London: John Murray Publishers Ltd.

Ilbery, B. and Bowler, I. (2014) From agricultural productivism to post-productivism. In Ilbery, B. (ed.) *The Geography of Rural Change*, pp. 57–84. London: Routledge. https://doi.org/10.4324/9781315842608

Karanasiou, A., Moreno, N., Moreno, T. et al. (2012) Health effects from Sahara dust episodes in Europe: Literature review and research gaps. *Environment International*, 47, pp. 107–114. https://doi.org/10.1016/j.envint.2012.06.012

Linares, C., Culqui, D., Belda, F. et al. (2021) Impact of environmental factors and Sahara dust intrusions on incidence and severity of COVID-19 disease in Spain. Effect in the first and second pandemic waves. *Environmental Science and Pollution Research*, 28(37), pp. 51948–51960. https://doi.org/10.1007/s11356-021-14228-3

Wang, Q., Gu, J. and Wang, X. (2020) The impact of Sahara dust on air quality and public health in European countries. *Atmospheric Environment*, 241, p. 117771. https://doi.org/10.1016/j.atmosenv.2020.117771

1. The Desert Biome of the World

Alonso, I. (2015) Where are we now and what is the gap? In Alonso, I., Underhill-Day, J. & Lake, S. (eds) *Proceedings of the 11th National Heathland Conference, 18–20 March 2015. Sunningdale Park, Berkshire*, pp. 79–86. York: Natural England.

Alonso, I, Sullivan, G. and Sherry, J. (2018) *Guidelines for the Selection of Biological SSSIs. Part 2: Detailed Guidelines for Habitats and Species Groups*. Peterborough: Joint Nature Conservation Committee.

Brandt, M., Tucker, C.J., Kariryaa, A. et al. (2020). An unexpectedly large count of trees in the West African Sahara and Sahel. *Nature* 587, pp. 78–82. https://doi.org/10.1038/s41586-020-2824-5

Bristow, C.S., Jol, H.M., Augustinus, P. et al. (2010). Slipfaceless 'whaleback' dunes in a polar desert, Victoria Valley, Antarctica: Insights from ground penetrating radar. *Geomorphology*, 114(3), pp. 361–372. https://doi.org/10.1016/j.geomorph.2009.08.001

Cary, S.C., McDonald, I.R., Barrett, J.E. and Cowan, D.A. (2010) On the rocks: the microbiology of Antarctic Dry Valley soils. *Nature Reviews Microbiology*, 8(2), pp. 129–138. https://doi.org/10.1038/nrmicro2281

Cherlet, M., Hutchinson, C., Reynolds, J. et al. (eds) *World Atlas of Desertification*. Luxembourg: Publication Office of the European Union. https://doi.org/10.2760/06292

Davies, J. et al. (2012) *Conserving Dryland Biodiversity*. Nairobi: International Union for Conservation of Nature.

De Smidt, J.T. (1977) Heathland vegetation in the Netherlands. *Phytocoenologia*, 4(3), pp. 258–316. https://doi.org/10.1127/phyto/4/1977/258

Díaz, J. et al. (2017) Saharan dust intrusions in Spain: Health impacts and associated synoptic conditions. *Environmental Research* 156, pp. 455–467. https://doi.org/10.1016/j.envres.2017.03.047

Diemont, W.H., Webb, N. and Degn, H.J. (1996) A pan European view on heathland conservation. In Anon. *Proceedings of the National Heathland Conference, 18–20 September 1996, Hampshire, UK*, pp. 21–32.

Ekaykin, A., Eberlein, L., Lipenkov, V. et al. (2016). Non-climatic signal in ice core records: lessons from Antarctic megadunes. *The Cryosphere* 10(3), pp. 1217–1227. https://doi.org/10.5194/tc-10-1217-2016

El Fadli, K.I. Cerveny, R.S., Burt, C.C., et al. (2013) World Meteorological Organization assessment of the purported world record 58 C temperature extreme at El Azizia, Libya (13 September 1922). *Bulletin of the American Meteorological Society*, 94(2), pp. 199–204. https://doi.org/10.1175/BAMS-D-12-00093.1

Engelstaedter, S., Tegen, I. and Washington, R. (2006) North African dust emissions and transport. *Earth-Science Reviews* 79(1–2), pp. 73–100. https://doi.org/10.1016/j.earscirev.2006.06.004

Ezcurra, E. (ed.) (2006) *Global Deserts Outlook*. Nairobi: UNEP.

Fahnestock, M.A., Scambos, T.A., Shuman, C.A. et al. (2000) Snow megadune fields on the East Antarctic Plateau: Extreme atmosphere-ice interaction. *Geophysical Research Letters*, 27(22), pp. 3719–3722. https://doi.org/10.1029/1999gl011248

Fagúndez, J. (2012) Heathlands Confronting global change: Drivers of biodiversity loss from past to future scenarios. *Annals of Botany*, 111(2), pp. 151–172. https://doi.org/10.1093/aob/mcs257

Farrell, L. (1993) *Lowland Heathland: The Extent of Habitat Change*. English Nature Science No. 12. Peterborough: English Nature.

Gaur, M.K. and Squires, V.R. (2018) Geographic extent and characteristics of the world's arid zones and their peoples. In Gaur, M.K. and Squires, V.R. (eds) *Climate Variability Impacts on Land Use and Livelihoods in Drylands*, pp. 3–20. Cham: Springer. https://doi.org/10.1007/978-3-319-56681-8

Goordial, J., Davila, A., Lacelle, D. et al. (2016). Nearing the cold-arid limits of microbial life in permafrost of an upper dry valley, Antarctica. *The ISME Journal*, 10(7), pp. 1613–1624. https://doi.org/10.1038/ismej.2015.239

Gray, M. (2018) Geodiversity. In Reynar, E. and Brilha, J. (eds) *Geoheritage: Assessment, Protection and Management*, pp. 13–25. Elsevier, Amsterdam. https://doi.org/10.1016/C2015-0-04543-9

Gray, M. (2019) Geodiversity, geoheritage and geoconservation for society. *International Journal of Geoheritage and Parks*, 7(4), pp. 226–236. https://doi.org/10.1016/j.ijgeop.2019.11.001

Harris, N. (2003) *Atlas of the World's Deserts*. New York: Fitzroy Dearborn. https://doi.org/10.4324/9780203491669

Hawley, G., Anderson, P., Gash, M., Smith, P., Higham, N., Alonso, I., Ede, J. and Holloway, J. (2008) *Impact of Heathland Restoration and Re-Creation Techniques on Soil Characteristics and the Historical Environment*. Natural England Research Reports, Number 010. York: Natural England.

HM Government (1994) *Biodiversity: The UK Action Plan*. CM 2428. London: HMSO.

Holzapfel, C. (2008) Deserts. In Jørgensen, S.E. and Fath, B.D. *Encyclopedia of Ecology*, pp. 879–898. Cambridge, MA: Academic Press.

Huang, J. and Hartemink, A.E. (2020) Soil and environmental issues in sandy soils. *Earth-Science Reviews* 208, 103295. https://doi.org/10.1016/j.earscirev.2020.103295

Karanasiou, A., Moreno, N., Moreno, T., Viana, M., de Leeuw, F. and Querol, X., 2012. Health effects from Sahara dust episodes in Europe: Literature review and research gaps. *Environment International*, 47, pp. 107–114. https://doi.org/10.1016/j.envint.2012.06.012

Kidron, G.J., Barzilay, E. and Sachs, E. (2000) Microclimate control upon sand microbiotic crusts, western Negev Desert, Israel. *Geomorphology*, 36(1–2), pp. 1–18. https://doi.org/10.1016/s0169-555x(00)00043-x

Linares, C., Culqui, D., Belda, F. et al. (2021) Impact of environmental factors and Sahara dust intrusions on incidence and severity of COVID-19 disease in Spain. Effect in the first and second pandemic waves. *Environmental Science and Pollution Research*, 28(37), pp. 51948–51960. https://doi.org/10.1007/s11356-021-14228-3

Millennium Ecosystem Assessment (2005) *Ecosystems and Human Well-being: Desertification Synthesis*. Washington, DC: World Resources Institute.

Olmeda, C., Šefferová, V., Underwood, E. et al. (eds) (2020) *EU Action Plan to Maintain and Restore to Favourable Conservation Status the Habitat Type 4030 European Dry Heaths*. Luxembourg: European Commission.

Rackham, O. (2020) *History of the Countryside*. London: Weidenfeld & Nicolson.

Rivas-Martínez, S., Penas, A. and Díaz, T. (2004) *Biogeographic Map of Europe*. León: University of León Cartographic Service.

Shrubsole, G. (2022). *The Lost Rainforests of Britain*. Glasgow: HarperCollins UK. https://doi.org/10.1016/s0262-4079(22)02185-6

Stolle, F. and Poole, J. (2017) We discovered 1.8 million square miles of forest in the desert, accessed at https://www.wri.org/blog/2017/05/we-discovered-18-million-square-miles-forest-desert (24 March 2021)

Tanaka, T.Y. and Chiba, M. (2006) A numerical study of the contributions of dust source regions to the global dust budget. *Global and Planetary Change*, 52(1–4), pp. 88–104. https://doi.org/10.1016/j.gloplacha.2006.02.002

Tomlin, S. (1999) Vast snow dunes frozen in time. *Nature*, 402(6764), pp. 860–860. https://doi.org/10.1038/47212

US Geological Survey (2011) *Sea Level and Climate: Fact Sheet fs002–00*. UGS: St. Petersburg, FL.

Wang, Q., Gu, J. and Wang, X. (2020) The impact of Sahara dust on air quality and public health in European countries. *Atmospheric Environment*, 241, p. 117771. https://doi.org/10.1016/j.atmosenv.2020.117771

White, R.P. and Nackoney, J. (2003) *Drylands, People, and Ecosystem Goods and Services: A Web-Based Geospatial Analysis*. Washington USA: World Resources Institute.

Whitford, W. (2002) *Ecology of Desert Systems*. San Diego: Academic Press.

Whitford, W.G. and Duval, B.D. (2019) *Ecology of Desert Systems*. Second Edition. Cambridge, Massachusetts: Academic Press https://doi.org/10.1016/C2017-0-02227-9

Zhao, Y., Norouzi, H., Azarderakhsh, M. et al. (2021). Global patterns of hottest, coldest, and extreme diurnal variability on Earth. *Bulletin of the American Meteorological Society*, 102(9), pp. E1672–E1681. https://doi.org/10.1175/bams-d-20-0325.1

2. Spain: Bardenas Reales – Iberian Badlands

Alagón, J.M. and Vázquez Astorga, M. (2015) Escuelas de "sabor agrario" en los pueblos creados por el Instituto Nacional de Colonización en la zona de la Violada-canal de Monegros I, Aragón. *Espacio, Tiempo y Educación* 2(1), pp. 281–308. https://doi.org/10.14516/ete.2015.002.001.014

Bardenas Reales de Navarra Natural Park (2017) *Espacio Natural Protegido: El Parque Natural, Reserva de la Biosfera, Zona Especial de Conservación*. Accessed at: https://bardenasreales.es/turismo/espacio-natural/ (31 August 2020)

Bardenas Reales de Navarra Natural Park (2019) Plano Nº 6_mapa de Habitats Prioritarios y de Interés Según Directiva Hábitats 92/43 Cee Para la Creación de la Red NATURA 2000. Accessed at: https://bardenasreales.es/wp-content/uploads/2019/11/plano6-habitats-prioritarios.pdf (21 August 2020)

Calvo-Cases, A. et al. (2014). Badlands in the Tabernas Basin, Betic chain. In Gutiérrez, F., Gutiérrez, M. (eds) *Landscapes and Landforms of Spain*, pp. 197–211. Springer, Dordrecht. https://doi.org/10.1007/978-94-017-8628-7_17

Comunidad de Bardenas Reales de Navarra (1998) *Plan de Ordenacion de los Recursos Naturales de Bardenas Reales de Navarra*. Accessed at: https://bardenasreales.es/wp-content/uploads/2019/11/plan-ordenacion-recursos-naturales.pdf (16 March 2021)

Desir, G., Marin, C., and Guerrero, J. (2005) Badlands and talus flatirons in the Bardenas Reales region. In Desir, G., Gutierrez, F., and Gutierrez, M. (eds) *Proceedings of the Sixth International Conference on Geomorphology, Zaragoza, Spain, 7 11 September 2005*, pp. 55–95. Kronos: Deerfield Beach, FL, USA.

Desir, G., Marín, C. and Guerrero, J. (2013) Caracterización de la erosión en áreas acarcavadas de la Fm. Tudela (Bardenas Reales, Navarra). *Cuadernos de Investigación Geográfica*, 35(2), pp. 195–213. https://doi.org/10.18172/cig.1218

European Environment Agency (2019) *Bardenas Reales*. Accessed at: https://eunis.eea.europa.eu/sites/ES2200037 (9 March 2023)

Froiland, S.G. (1990) *Natural History of the Black Hills and Badlands*. Sioux Falls: The Centre for Western Studies.

Genova, A. (2018) Thousands of People Live in These Ancient Spanish Caves. *National Geographic* magazine. Accessed at: https://www.nationalgeographic.com/culture/article/cave-underground-dwellers-ancient-modern-granada-spain (21 August 2020)

Mcmahon, B.J., Giralt, D., Raurell, M. et al. (2010). Identifying set-aside features for bird conservation and management in northeast Iberian pseudo-steppes. *Bird Study*, 57(3), pp. 289–300. https://doi.org/10.1080/00063651003749680

P.O.R.N de la Comunidad de Bardenas Reales de Navarra (2019) *Proyecto Conejo. El Centro Cinegético*. Accessed at: https://bardenasreales.es/wp-content/uploads/2019/08/Proyecto-Conejo-El-Centro-Cinegetico.pdf (3 June 2021)

Samanes, A.F. (1949) Una descripción de las Bardenas Reales en el siglo XVIII. *Principe De Viana*, 10(37), pp. 475–481. https://doi.org/10.35462/pv

Weiner, H.R. (1978) New Communities in Franco Spain: The rural towns of the "Instituto Nacional de Colonizacion". *Urbanism Past & Present*, 7, pp. 13–20.

3. The Netherlands: Dry Wilderness in a Watery Country

Blanken, M.C. (1976) The Dutch "Miracle". In: *Force of Order and Methods: An American view into the Dutch Directed Society*. Studies in Social Life, 19. Dordrecht: Springer. https://doi.org/10.1007/978-94-015-0638-0_1

Blume, H-P, Leinweber, P. (2004) Plaggen soils: landscape history, properties, and classification. *Journal of Plant Nutrition and Soil Science*, 167(3). https://10.1002/jpln.200420905

Brakensiek, S. (2002) The management of common land in north-western Germany. In Moor, M. de, Shaw-Taylor, L. and Warde, P. (eds) (2002) *The Management of Common Land in North West Europe, c.1500–1850*, pp. 225–245. Turnhout: Brepols Publishers. https://doi.org/10.1484/M.CORN-EB.4.00180

Castel, I., Koster, E. and Slotboom, R. (1989) Morphogenetic aspects and age of Late Holocene eolian drift sands in Northwest Europe. *Zeitschrift für Geomorphologie*, 33(1), pp. 1–26. https://doi.org/10.1127/zfg/33/1989/1

Davidson, D., Dercon, G., Simpson, I. et al. (2007) The identification and significance of inputs to anthrosols in North-West Europe. *Atti della Societa Toscana di Scienze Naturali-Memorie Serie A*, 112, pp. 79–83.

De Glopper, R.J. and Smits, H. (1974) Reclamation of land from the sea and lakes in the Netherlands. *Outlook on Agriculture*, 8(3), pp. 148–155. https://doi.org/10.1177/003072707400800

De Keyzer, M. (2016) All we are is dust in the wind: The social causes of a "subculture of coping" in the late medieval coversand belt. *Journal for the History of Environment and Society*, 1(1), pp. 1–35. https://doi.org/10.1484/J.JHES.5.110827

De Keyzer, M. (2019) Common challenges, different fates. The causal factors of failure or success in the commons: The pre-modern Brecklands (England) and the Campine (Southern Low Countries) compared. In Haller, T., Breu, T., De Moor, T. et al. (eds) *The Commons in a Glocal World: Global Connections and Local Responses*, pp. 123–140. London: Routledge. https://doi.org/10.4324/9781351050982

De Keyzer, M. and Bateman, M.D. (2018) Late Holocene landscape instability in the Breckland (England) drift sands. *Geomorphology*, 323, pp. 123–134. https://doi.org/10.1016/j.geomorph.2018.06.014

Fanta, J. and Siepel, H. (eds) (2010) *Inland Drift Sand Landscapes: Origin and History; Relief, Forest and Soil Development; Dynamics and Management*. Zeist: KNNV Publishing.

Fremantle, K. (1970) A visit to the United Provinces and Cleves in the time of William III: Described in Edward Southwell's journal. *Netherlands Yearbook for History of Art*, 21, pp. 39–68. http://www.jstor.org/stable/43875636

Giani L., Makowsky L. and Mueller, K. (2014) Plaggic Anthrosol: Soil of the Year 2013 in Germany. An overview on its formation, distribution, classification, soil function and threats. In: *Journal of Plant Nutrition and Soil Science* (177) Heft 3, S. pp. 320–329. https://doi.org/10.1002/jpln.201300197

Grinberg, D.I. (1977) *Housing in the Netherlands 1900–1940*. Delft: Delft University Press. https://doi.org/10.1007/978-94-011-6459-7

Groenewoudt, B.J. (2009). An exhausted landscape. Medieval use of moors, mires and commons in the Eastern Netherlands. *Ruralia*, pp. 149–180. https://doi.org/10.1484/m.ruralia-eb.3.1169

Groenman-Van Waateringe, W. (1996) Wasteland: A buffer in the medieval economy, *Actes des Congrès de la Société d'Archéologie Médiévale*, 5(1), pp. 113–117.

Heidinga, H.A. (1984) Indications of severe drought during to 10th century AD from an inland dune area in the central Netherlands. *Geologie en Mijnbouw Delft*, 63(3), pp. 241–248.

Heidinga, H.A. (2010) The birth of a desert: The Kootwijkerzand. In Fanta, J. and Siepel, H. (eds) *Inland Drift Sand Landscapes: Origin and History; Relief, Forest and Soil Development; Dynamics and Management*, pp. 65–79. Zeist: KNNV Publishing.

Huisman, D.J. and Milek, K. (2017) Turf as Construction Material, chapter in book: Dr. Cristiano Nicosia, Em. Prof. Dr. Georges Stoops (2017) Archaeological Soil and Sediment Micromorphology. pp. 113–119. John Wiley & Sons. https://doi.org/10.1002/9781118941065

Jones, A., Montanarella, L., Jones, R. et al. (2005) *Soil Atlas of Europe*. Luxembourg: European Commission Office for Official Publications of The European Communities.

Jungerius, P.D. and Riksen, M.J. (2010) Contribution of laser altimetry images to the geomorphology of the Late Holocene inland drift sands of the European Sand Belt. *Baltica*, 23(1), 59–70.

BIBLIOGRAPHY

Ketner-Oostra, R. and Sýkora, K.V. (2009) Vegetation change in a lichen-rich inland drift sand area in the Netherlands. *Phytocoenologia*, 38(4), pp. 267–286. https://doi.org/10.1127/0340-269x/2008/0038-0267

Koers, A. (2019) *Ommerschans Bedelaarsgesticht*. At: https://hethistorischportaal.nl/wp-content/uploads/OMMERSCHANS.pdf

Koster, E.A., 2009. The "European Aeolian Sand Belt": Geoconservation of Drift Sand Landscapes. *Geoheritage*, 1(2), pp. 93–110. https://doi.org/10.1007/s12371-009-0007-8

Lascaris, M., 1999. Zandverstuivingen op de noordwestelijke Veluwe (Shifting sands of the north-western Veluwe region). *Historisch Geografisch Tijdschrift*, 17, pp. 54–63.

Meulen, F., Hagen, H.G. and Kruijsen, B. (1987) *Campylopus introflexus*: Invasion of a moss in Dutch coastal dunes. In *Proceedings of the Koninklijke Nederlandse Akademie van Wetenschappen*. Series C: Biological and Medical Sciences, 90, pp. 73–80.

Neefjes, J. (2018) *Landschapsbiografie van de Veluwe: Historisch-Landschappelijke Karakteristieken en Hun Ontstaan*. Amersfoort: Rijksdienst voor het Cultureel Erfgoed.

Pape, J.C. (1970) Plaggen soils in the Netherlands. *Geoderma*, 4(3), pp. 229–255. https://doi.org/10.1016/0016-7061(70)90005-4

Pedroli, B., van Doorn, A., de Blust, G. et al. (2007) *Europe's Living Landscapes: Essays Exploring our Identity in the Countryside*. Zeist, the Netherlands: KNNV Publishing.

Riksen, M.J. and Goossens, D. (2007) The role of wind and splash erosion in inland drift-sand areas in the Netherlands. *Geomorphology*, 88(1–2), pp. 179–192. https://doi.org/10.1016/j.geomorph.2006.11.002

Riksen, M.J., Goossens, D. and Jungerius, P.D. (2004) The role of wind and water erosion in drift-sand areas in The Netherlands. In Visser, S.M. and Cornells, W.M. (eds) *Wind and Rain Interaction in Erosion*, pp. 97–129. Wageningen: Wageningen University.

Riksen, M., Ketner-Oostra, R., van Turnhout, C. et al. (2006) Will we lose the last active inland drift sands of western Europe? The origin and development of the inland drift-sand ecotype in the Netherlands. *Landscape Ecology*, 21(3), pp. 431–447. https://doi.org/10.1007/s10980-005-2895-6

Riksen, M.J., Spaan, W.P. and Stroosnijder, L. (2008) How to use wind erosion to restore and maintain the inland drift-sand ecotype in the Netherlands? *Journal for Nature Conservation*, 16(1), pp. 26–43. https://doi.org/10.1016/j.jnc.2007.07.002

Roymans, N. and Kluiving, S. (2012) Soil degradation and shifting habitation patterns in the sand landscapes of the Southern Netherlands. In Bebermeier, W., Hebenstreit, R., Kaiser, E. et al. (eds) Landscape Archaeology. Proceedings of the International Conference held in Berlin, 6th–8th June 2012. *Journal for Ancient Studies*. Special Volume 3, pp. 47–53. https://doi.org/10.17171/5-3

Schama, S. (1995) *Landscape and Memory*. HarperCollins, London. https://doi.org/10.1177/147447409600300309

Schrauwers, A. (2001) The "benevolent" colonies of Johannes van den Bosch: continuities in the administration of poverty in the Netherlands and Indonesia. *Comparative Studies in Society and History*, 43(2), pp. 298–328. https://doi.org/10.1017/s0010417501003504

Snijders, R. (1984) De Veluwezoom. Zutphen: Terra, quoted in Arts, K., Fischer, A. and van der Wal, R. (2011) Wilderness – Between the promise of hell and paradise: A cultural-historical exploration of a Dutch National Park. In: Watson, A, Murrieta-Saldivar, J., McBride, B. *Science and Stewardship to Protect and Sustain Wilderness Values: Ninth World Wilderness Congress Symposium, November 6–13, 2009, Merida, Yucatan, Mexico*. Proceedings RMRS-P-64. Fort Collins, CO: US Department of Agriculture, Forest Service, Rocky Mountain Research Station, pp. 118–124.

Spaan, W.P., Riksen, M.J. and Winteraeken, H.J. (2006) Dutch policy and practices: then and now, *Archives of Agronomy and Soil Science*, 52(2), pp. 233–241. https://doi.org/10.1080/03650340600603861

Sparrius, L.B. and Kooijman, A.M. (2011) Invasiveness of *Campylopus introflexus* in drift sands depends on nitrogen deposition and soil organic matter. *Applied Vegetation Science*, 14(2), pp. 221–229. https://doi.org/10.1111/j.1654-109x.2010.01120.x

Sparrius, L. and Riksen, M. (2019) *Evaluatie van elf Jaar Stuifzandbeheer op de Veluwe 2007–2018 (No. 23)*. Bryologische and Lichenologische Werkgroep. Accessed at: https://edepot.wur.nl/507822 (28 November 2022)

Stichting Het Nationale Park De Hoge Veluwe Foundation (2021) *General Policy Plan 2020–2025: International Appeal*. Accessed at: https://www.hogeveluwe.nl/files/2021-3981%20Hoge%20Veluwe%20Beleidsplan%202020-2025%20(Engels%20def)%20(002).pdf (20 October 2022)

Stuit, H. (2020) Dutch domestic colonization: From rural idyll to prison museum. *Collateral: Online Journal for Cross-Cultural Close Reading*, 23. Accessed at: http://www.collateral-journal.com/index.php?cluster=23 (7 April 2023)

Turnhout, E., Hisschemöller, M. and Eijsackers, H. (2004) The role of views of nature in Dutch nature conservation: The case of the creation of a drift sand area in the Hoge Veluwe National Park. *Environmental Values*, 13(2), pp. 187–198. https://doi.org/10.3197/0963271041159868

Urbanski, L., Schad, P., Kalbitz, K. et al. (2022) Legacy of plaggen agriculture: High soil organic carbon stocks as result from high carbon input and volume increase. *Geoderma*, 406, p. 115513. https://doi.org/10.1016/j.geoderma.2021.115513

Van Huijgevoort, M.H., Voortman, B.R., Rijpkema, S. et al. (2020) Influence of climate and land use change on the groundwater system of the Veluwe, The Netherlands: A historical and future perspective. *Water (Switzerland)* 12, pp. 1–16. https://doi.org/10.3390/w12102866

Van Nederveen Meerkerk, E. (2016) Grammar of difference? The Dutch colonial state, labour policies, and social norms on work and gender, c.1800–1940. *International Review of Social History*, 61(S24), pp. 137–164. https://doi.org/10.1017/s0020859016000481

Van Turnhout, C. (2005) The disappearance of the Tawny Pipit *Anthus campestris* as a breeding bird from The Netherlands and Northwest-Europe. *Limosa*, 78, pp. 1–14.

Ward, C. (2002) *Cotters and Squatters: Housings Hidden History*. Nottingham: Five Leaves Publications.

Webb, N.R. (1998) The traditional management of European heathlands. *Journal of Applied Ecology*, 35(6), pp. 987–990.

Wigman, A.B. (1910) Hent uut 't Zaand. *Buiten* magazine, 18 June 1910. Accessed at: https://www.de-veluwenaar.nl/2013/09/17/hent-uut-t-zaandde-laatste-bewoners-van-een-plaggenhut/ (28 June 2022)

Woestenburg, M. (2018) Heathland farm as a new commons? *Landscape Research*, 43(8), pp. 1045–1055. https://doi.org/10.1080/01426397.2018.1503236

4. France: The Desert That Disappeared

Aldhuy, J. (2007) La transformation des Landes de Gascogne, de la mise en valeur comme colonisation intérieure (XVIIIe–XIXe siècles)? *Sud-Ouest Européen*, 23(1), pp. 17–28. https://doi.org/10.3406/rgpso.2007.2935

Alethea, E. (1855) *Pignadar: Or Three Days' Wanderings in the Landes*. London: Longman, Brown, Green and Longmans.

Bertran, P., Bateman, M.D., Hernandez, M. et al. (2011) Inland aeolian deposits of south-west France: facies, stratigraphy and chronology: inland aeolian deposits of SW France. *Journal of Quaternary Science*, 26(4), pp. 374–388. https://doi.org/10.1002/jqs.1461

Bertran, P., Andrieux, E., Bateman, M.D. et al. (2020) Mapping and chronology of coversands and dunes from the Aquitaine basin, southwest France. *Aeolian Research*, 47, p. 100628. https://doi.org/10.1016/j.aeolia.2020.100628

Blanchard, W.O. (1926) The Landes: Reclaimed Waste Lands of France. *Economic Geography* 2(2), pp. 249–255. https://doi.org/10.2307/140867

Bory de Saint-Vincent, M. (1826) *Dictionnaire Classique d'Histoire Naturelle*. Paris: Rey et Gravier.

Clarke, M., and Rendell, H.M. (2014) 'This restless enemy of all fertility': exploring paradigms of coastal dune management in western Europe over the last 700 years. *Transactions of the Institute of British Geographers*, 40(3), pp. 414–429. https://doi.org/10.1111/tran.12067

Cobb, C. (1910) The Landes and dunes of Gascony. *Journal of the Elisha Mitchell Scientific Society*, 26 (3), pp. 82–89.

Dickerson, J.W.T. and Wiryanti, J. (1978) Pellagra and mental disturbance. *Proceedings of the Nutrition Society*, 37(2), pp. 167–171. https://doi.org/10.1079/pns19780021

Espinosa, J., Palheiro, P., Loureiro, C. et al. (2019) Fire-severity mitigation by prescribed burning assessed from fire-treatment encounters in maritime pine stands. *Canadian Journal of Forest Research*, 49(2), pp. 205–211. https://doi.org/10.1139/cjfr-2018-0263

Fernandes, P.M. and Rigolot, E. (2007) The fire ecology and management of maritime pine (*Pinus pinaster* Ait.). *Forest Ecology and Management*, 241(1–3), pp. 1–13. https://doi.org/10.1016/j.foreco.2007.01.010

Flamichon, E. (1816) *Théorie de la Terre*. Pau: Chez Tonnet.

Fryberger, S.G., Krystinik, L.F. and Schenk, C.J. (1990) *Modern and Ancient Eolian Deposits: Petroleum Exploration and Production*. Denver: Rocky Mountain Section (SEPM).

Grasset de Saint-Sauveur, J. (1798) *Voyage à Bordeaux et dans Les Landes où sont décrits les moeurs, les usages et costumes du pays*. Paris: Pigoreau.

Hegyi, J., Schwartz, R.A. and Hegyi, V. (2004) Pellagra: dermatitis, dementia, and diarrhea. *International Journal of Dermatology*, 43(1), pp. 1–5. https://doi.org/10.1111/j.1365-4632.2004.01959.x

Institut National de L'Information Géographique et Forestière (2013) *Sylvoécorégion F 21: Landes de Gascognes*. Paris: Institut National de L'Information Géographique et Forestière.
Jolivet, C., Augusto, I., Trichet, P. and Arrouays, D. (2007) Les sols du massif forestier des Landes de Gascogne: Formation, histoire, propriétés et variabilité spatiale. *Revue Forestière Française*, (1). https://doi.org/10.4267/2042/8480
Krasnodębski, M. (2019). Challenging the pine: Epistemic underpinnings of techno-environmental inertia. *Journal for the History of Environment and Society*, 4, pp. 41–69. https://doi.org/10.1484/j.jhes.5.120675
Mauriac, F. (1927) *Thérèse Desqueyroux*. Paris: Éditions Grasset.
Métivier, Charles-Gabriel-François-Hyacinthe-Denis (1839) *De l'agriculture et du défrichement des Landes*. Bordeaux: Th. Lafargue.
Papy, L. (1947) L'ancienne vie pastorale dans la Grande Lande. *Revue Geographique des Pyrenees et du Sud-Ouest*, 18(1), pp. 5–16. https://doi.org/10.3406/rgpso.1947.1232
Ribadieu, H. (1859) *A Trip to the Bassin d'Arcachon*. Paris: Tardieu.
Ribadieu, H. and Dupuy, J. (1853) *Histoire de Bordeaux Pendant Le Règne de Louis XVI*. Bordeaux: Chez les Principaux Libraires.
Ribéreau-Gayon, M.D. (2000) Chemins et regards croisés dans les Landes de Gascogne: du XVIIIe siècle à nos jours. Le Monde alpin et rhodanien. *Revue Régionale d'Ethnologie*, 28(1), pp. 175–196.
Richer-de-Forges, A.C., Saby, N.P., Mulder, V.L., Laroche, B. and Arrouays, D. (2017) Probability mapping of iron pan presence in sandy podzols in South-West France, using digital soil mapping. *Geoderma Regional*, 9, pp. 39–46. https://doi.org/10.1016/j.geodrs.2016.12.005
Semba, R.D. (2000) Theophile Roussel and the elimination of Pellagra from 19th century France. *Nutrition*, 3(16), pp. 231–233. https://doi.org/10.1016/s0899-9007(99)00273-7
Sutton, K. (1977) Reclamation of wasteland during the 18th and 19th centuries. In Clout, H.D. (ed.) *Themes in the Historical Geography of France*, pp. 247–300. London: Academic Press.
Tastu, A. (1878) *Voyage en France*. Tours: Ad. Mame et Cie.
Temple, S. (2009) The natures of nation: Negotiating modernity in the Landes de Gascogne. *French Historical Studies*, 32(3), pp. 419–446.
Thore, J. (1810) *Promenade sur les côtes du Golfe de Gascogne*. Bordeaux: Brossier.
Timbal, J. and Maizeret, C. (1998) Plant biodiversity and sustainable management of the maritime pine forests of the Landes: Present situation and future prospects. *Revue Forestiere Francaise*, 50(5), pp. 403–424.
Traimond, B. (1986) Le voyage dans les Landes de Gascogne ou la traversée du Sahara français. *Études rurales*, n° 103–104, pp. 221–234.
Trichet, P., Jolivet, C., Arrouays, D. et al. (1999) Soil sustainable management for the maritime pine sylviculture in southwest France. *Etude et Gestion des Sols*, 6, pp. 197–214.
Whitlock Rose, E. (1906) *Cathedrals and Cloisters in the South of France: Vol. II*. New York: GP Putnam's Sons.

5. Pustynia Błędowska: The Polish Sahara

Adamczyk, A.F. (1990) Wpływ górnictwa rud cynku i ołowiu w rejonie olkuskim na wody podziemne i powierzchniowe. Zeszyty Naukowe Akademii Górniczo-Hutniczej w Krakowie. *Sozologia i Sozotechnika*, 32, pp. 41–55.
Adamska, E., Deptuła, M., Filbrandt-Czaja, A. et al. (2015) *Heathlands and Associated Communities in Kujawy and Pomerania: Management, Treatment and Conservation*. Toruń: Towarzystwo Naukowe w Toruniu.
Cabała, J. and Rahmonov, O. (2004) Cyanophyta and algae as an important component of biological crust from the Pustynia Błędowska Desert (Poland). *Polish Botanical Journal*, 49(1), pp. 93–100.
Całbecki, P. (2015) *The 14th European Heathlands Workshop: Heathlands of Protected and Military Training Areas in Northern Poland*. Torun: Nicolaus Copernicus University.
Caputa, Z. (2007) Diversity of albedo and longwave exchange and radiative efficiency coefficients on Bledow Desert area. *Pamietnik Pulawski*, 144, pp. 35–44.
Caputa, Z. (2016) The structure of the radiation balance on a sandy surface: Case the Błędów desert, Silesian Upland. *Ekológia (Bratislava)*, 35(2), pp. 114–125. https://doi.org/10.1515/eko-2016-0009
Chadýšiene, R. and Girgždys, A. (2008) Ultraviolet radiation albedo of natural surfaces. *Journal of Environmental Engineering and Landscape Management*, 16(2), pp. 83–88. https://doi.org/10.3846/1648-6897.2008.16.83-88
Czaja, S. (2001) Mining and hydrological transformations in Upper Silesia from the fifteenth to the nineteenth century. *The Geographical Journal*, 167(1), pp. 57–71. https://doi.org/10.1111/1475-4959.t01-1-00005

Dziechciarz, O. (2016, August 12) Błędowska Desert. *Olkuski Przeglad*. Accessed at https://przeglad.olkuski.pl/pustynia-bledowska/ (1 March 2023)

Izmailow, B. and Morkūnaitė, R. (2004) Problems of continental dune research and their solution methods in Poland. *Geografijos Metraštis*, 37(1–2).

Kai, H. and Iba, K. (2014) Temperature stress in plants. *Els* (Wiley Online Library). https://doi.org/10.1002/9780470015902.a0001320.pub2

Łabęcki, H.H. (1841) Wiadomość bibliograficzna o górnictwie w Polsce i naukach przyrodniczych ścisły związek z nimi mających. *Biblioteka Warszawska*, t. 4, s. 99–136.

Lech, K. and Jarosz, M. (2016) Identification of Polish cochineal (*Porphyrophora polonica* L.) in historical textiles by high-performance liquid chromatography coupled with spectrophotometric and tandem mass spectrometric detection. *Analytical and Bioanalytical Chemistry*, 408(12), pp. 3349–3358. https://doi.org/10.1007/s00216-016-9408-0

Mazur, E., Wojciech, M. and Kępa, K. (2017) *Complex Protection of Non-Forested Natural Habitats in Military Areas in NATURA 2000 Area: The Błędowska Desert*. LIFE12 NAT/PL/000031 After LIFE Conservation Plan. Accessed at: http://rzi-life-pustynia.pl (23 July 2022)

Motyka, J., and Postawa, A. (2013) Impact of Zn–Pb mining in the Olkusz ore district on the Permian aquifer (SW Poland). *Environmental Science and Pollution Research*, 20, 7582–7589. https://doi.org/10.1007/s11356-013-1740-8

Nałkowski, N. (1887) *Geograficzny rzut oka na dawną Polskę*. Warszawa.

Noy-Meir, I. (1973) Desert ecosystems: environment and producers. *Annual Review of Ecology and Systematics*, 4, pp. 25–51. https://doi.org/10.1146/annurev.es.04.110173.000325

Pełka-Gościniak, J. (2013) Human activity and aeolian relief of Starczynów "Desert", Poland. *Environmental & Socio-Economic Studies*, 1(3), pp. 1–6. https://doi.org/10.1515/environ-2015-0013

Pełka-Gościniak, J., Rahmonov, O. and Szczypek, T. (2014) The Błędów Desert: Past and future of the largest blow sands area in Poland. In Dulias, R. and Prokop, P. *Land Degradation and Reclamation in the Silesian Upland and the Polish Carpathians*. Sosnowiec: University of Silesia. Faculty of Earth Sciences.

Piech, K. (1924) Miraże w Pustyni Błędowskiej. *Kosmos*, 49, pp. 876–878.

Rahmonov, O. (2016) Pustynia Błędowska – fenomen krajobrazu. In Rybak A., Wójcik A.J. and Woźnicka, Z. (eds) *Dąbrowa Górnicza*, pp. 293–317. Wroclaw: Muzeum Miejskie.

Rahmonov, O. and Oles, W. (2010) Vegetation succession over an area of a medieval ecological disaster. The case of the Bledów Desert, Poland. *Erdkunde*, 64(3), pp. 241–255. https://doi.org/10.3112/erdkunde.2010.03.03

Rahmonow. O. (1999) *Procesy Zarastania Pustyni Błędowskiej*. Sosnowiec: Wydział Nauk o Ziemi Uniwersytetu Śląskiego.

Rzeczycki, T. (2013, 27 March) Legenda pustyni nieprawdziwej (cz. 1). *Interia Wydarzenia*. Accessed at https://wydarzenia.interia.pl/historia/news-legenda-pustyni-nieprawdziwej-cz-1,nId,947088 (3 March 2023)

Scheib, A.J., Birke, M., and Dinelli, E. (2014) Geochemical evidence of aeolian deposits in European soils. *Boreas*, 43, pp. 175–192. https://doi.org/10.1111/bor.12029

Schmidt-Przewoźna, K. (2018) History of red colours in Poland: Polish Cochineal *Porophyropora polonica* L. and other natural dyes. In: Godyń, M., Groborz, B. and Kwiatkowska-Lubańska, A. (eds) *Colour-Culture-Science*, pp. 34–41. Kraków: Faculty of Industrial Design, Jan Matejko Academy of Fine Arts in Krakow.

Sielezniew, M., Dziekańska, I. and Stankiewicz-Fiedurek, A.M. (2009) Multiple host-ant use by the predatory social parasite *Phengaris* (=*Maculinea*) *arion* (Lepidoptera, Lycaenidae). *Journal of Insect Conservation*, 14(2), pp. 141–149. https://doi.org/10.1007/s10841-009-9235-0

Sznajder, M.J. (2017) Case study: The submetropolitan Błędów Desert, an ambivalent approach to devastation of the natural environment. In Sznajder, M.J. (ed.) *Metropolitan Commuter Belt Tourism*, pp. 49–51. Abingdon-on-Thames, UK: Routledge.

Szostak, M. et al. (2016) Monitoring the secondary forest succession and land cover/use changes of the Błędów Desert (Poland) using geospatial analyses. *Quaestiones Geographicae*, 35(3), pp. 1–13. https://doi.org/10.1515/quageo-2016-0022

Tyc, A., Czylok, A. and Rahmonov. O. (1999) Human impact and spontaneous regeneration of a karst-aeolian ecosystem in an anthropogenic desert near Olkusz (Silesian Upland, Poland). In Barany-Kevei, I. and Gunn, J. (eds) *Essays in the Ecology and Conservation of Karst. Acta Universitatis Szegediensis*, 36, pp. 70–77.

Walnik, A. (2014) *Layman Report: Project: Active protection of a complex of priority of on-sand habitats on the Bledowska Desert*. Klucze: Klucze Poligrafii.

Wang, B., Ye, T., Li, C. et al. (2022) Cell damage repair mechanism in a desert green algae Chlorella sp. against UV-B radiation. *Ecotoxicology and Environmental Safety*, 242, p. 113916. https://doi.org/10.1016/j.ecoenv.2022.113916

Whitford, W. (2002) *Ecology of Desert Systems*. San Diego: Academic Press.

Wolf, L., Rizzini, L., Stracke, R. et al. (2010) The Molecular and physiological responses of Physcomitrella patens to ultraviolet-B radiation. *Plant Physiology*, 153(3), pp. 1123–1134. https://doi.org/10.1104/pp.110.154658

6. Iceland: Cold Black Deserts

Alho, P. (2003) Land cover characteristics in NE Iceland with special reference to jökulhlaup geomorphology. *Geografiska Annaler: Series A, Physical Geography*, 85(3–4), pp. 213–227. https://doi.org/10.1111/j.0435-3676.2003.00201.x

Aradóttir, Á.L., Petursdottir, T., Halldorsson, G. et al. (2013) Drivers of ecological restoration: Lessons from a century of restoration in Iceland. *Ecology and Society*, 18(4). https://doi.org/10.5751/ES-05946-180433

Arnalds, A. (1987) Ecosystem disturbance in Iceland. *Arctic and Alpine Research*, 19(4), pp. 508–513.

Arnalds, A. (2005) Approaches to landcare – a century of soil conservation in Iceland. *Land Degradation and Development*, 16(2), pp. 113–125.

Arnalds, O. (1999) Soil survey and databases in Iceland. *European Soil Bureau Research Report No. 6*, pp. 91–96.

Arnalds, O. (2008) Soils of Iceland. *Jökull*. no. 58, pp. 409–421.

Arnalds, O. (2010) Dust sources and deposition of aeolian materials in Iceland. *Icelandic Agricultural Sciences*, 23(1), pp. 3–21.

Arnalds, O. (2015) *The Soils of Iceland*. Dordrecht: Springer. https://doi.org/10.1007/978-94-017-9621-7

Arnalds, O. (2019) Development of perverse environmental subsides for sheep production in Iceland. *Agricultural Sciences*, 10(9), pp. 1135–1151. https://doi.10.4236/as.2019.109086

Arnalds, O. and Barkarson, B.H. (2003) Soil erosion and land use policy in Iceland in relation to sheep grazing and government subsidies. *Environmental Science & Policy*, 6(1), pp. 105–113. https://doi.org/10.1016/s1462-9011(02)00115-6

Arnalds, O. and Kimble, J. (2001) Andisols of deserts in Iceland, *Soil Science Society of America Journal*, 65(6), pp. 1778–1786. https://doi.org/10.2136/sssaj2001.1778

Arnalds, O., Gisladottir, F.O., & Sigurjonsson, H. (2001). Sandy deserts of Iceland: an overview. *Journal of Arid Environments*, 47(3), 359–371. https://doi.org/10.1006/jare.2000.0680

Arnalds, O., Thorarinsdottir, E.F., Metusalemsson, S. et al. (2001) *Soil Erosion in Iceland*. Hella: Soil Conservation Service.

Arnalds, O., Thorarinsdottir, E., Thorsson, J. et al. (2013) An extreme wind erosion event of the fresh Eyjafjallajökull 2010 volcanic ash. *Scientific Reports*, 3(1), p. 1257. https://doi.org/10.1038/srep01257

Arnalds, O., Olafsson, H. and Dagsson-Waldhauserova, P. (2014) Quantification of iron-rich volcanogenic dust emissions and deposition over ocean from Icelandic dust sources. *Biogeosciences*, 11, 5941–5967. https://doi.org/10.5194/bgd-11-5941-2014

Arnalds, O., Dagsson-Waldhauserova, P. and Olafsson, H. (2016) The Icelandic volcanic aeolian environment: Processes and impacts – A review. *Aeolian Research*, 20, pp. 176–195.

Arnalds, O., Dagsson-Waldhauserova, P. and Olafsson, H. (2018) Dyngjusandur; A rapidly evolving hyperactive dust source north of Vatnajökull glacier, Iceland. *Geophysical Research Abstracts*, 20, p. 14764. Proceedings from the EGU General Assembly conference, 4–13 April, 2018, Vienna, Austria.

Baratoux, D. et al. (2011) Volcanic sands of Iceland – Diverse origins of aeolian sand deposits revealed at Dyngjusandur and Lambahraun: Volcanic sands of Iceland. *Earth Surface Processes and Landforms*, 36(13), pp. 1789–1808. https://doi.org/10.1002/esp.2201

Barrio, I.C. and Arnalds, Ó. (2022) Agricultural Land Degradation in Iceland. In: Pereira, P., Muñoz-Rojas, M., Bogunovic, I. et al. (eds) *Impact of Agriculture on Soil Degradation II. The Handbook of Environmental Chemistry*, pp. 159–177. Cham: Springer. https://doi.org/10.1007/978-3-031-32052-1

Benediktsson, K. (2015). Floral hazards: Nootka lupin in Iceland and the complex politics of invasive life. *Geografiska Annaler: Series B, Human Geography*, 97(2), pp. 139–154. https://doi.org/10.1111/geob.12070

Browne, J.R. (1867) *The Land of Thor*. New York: Harper & Brothers.

Crofts, R. (2011) *Healing the Land: The Story of Land Reclamation and Soil Conservation in Iceland*. Hella: Soil Conservation Service of Iceland.

Dagsson-Waldhauserova, P., Arnalds, O. and Olafsson, H. (2014) Long-term variability of dust events in Iceland (1949–2011). *Atmospheric Chemistry and Physics*, 14(24), pp. 13411–13422. https://doi.org/10.5194/acp-14-13411-2014

Dagsson-Waldhauserova, P., Arnalds, O., Olafsson, H. et al. (2015). Snow–dust storm: Unique case study from Iceland, March 6–7, 2013. *Aeolian Research*, 16, pp. 69–74. https://doi.org/10.1016/j.aeolia.2014.11.001

Dagsson-Waldhauserova, P., Magnusdottir, A., Olafsson, H. et al. (2016) The spatial variation of dust particulate matter concentrations during two Icelandic dust storms in 2015. *Atmosphere*, 7(6), p. 77. https://doi.org/10.3390/atmos7060077

Dagsson-Waldhauserova, P., Arnalds, O. and Olafsson, H. (2017) Long-term dust aerosol production from natural sources in Iceland. *Journal of the Air & Waste Management Association*, 67(2), pp. 173–181. https://doi.org/10.1080/10962247.2013.805703

Davies, S.M., Larsen, G., Wastegård, S. et al. (2010) Widespread dispersal of Icelandic tephra: how does the Eyjafjöll eruption of 2010 compare to past Icelandic events? *Journal of Quaternary Science*, 25(5), pp. 605–611. https://doi.org/10.1002/jqs.1421

Diamond, J. (2005) *Collapse: How Societies Choose to Fail or Succeed*. New York: Viking. https://doi.org/10.62608/2158-0669.1017

Eddudóttir, S.D., Erlendsson, E. and Gísladóttir, G. (2020) Landscape change in the Icelandic highland: A long-term record of the impacts of land use, climate and volcanism, *Quaternary Science Reviews*, 240(106363), p. 106363. https://doi.org/10.1016/j.quascirev.2020.106363

Elíasson, J., Kjaran, S.P., Holm, S.L., Gudmundsson, M.T. and Larsen, G. (2007) Large hazardous floods as translatory waves. *Environmental Modelling & Software*, 22(10), pp. 1392–1399. In Olafur Arnalds (2015) The Soils of Iceland. Dordrecht: Springer. https://doi.org/10.1007/978-94-017-9621-7

Gudmundsson, G. (2023) Letter from Iceland. *Respirology*, 28(4), pp. 404–405.

Gudmundsson, G. (2010) Respiratory health effects of volcanic ash with special reference to Iceland. A review. *The Clinical Respiratory Journal*, 5(1), pp. 2–9. https://doi.org/10.1111/j.1752-699x.2010.00231.x

Hallsdóttir, B.S., Harðardóttir, K., Guðmundsson, J. et al. (2009) *National Inventory Report Iceland 2009: Submitted under the United Nations Framework Convention on Climate Change*. Reykjavík: Environment Agency of Iceland.

Hemond, C., Arndt, N.T., Lichtenstein, U., Albrecht, W.F. et al. (1993) The Heterogeneous Iceland Plume. *Journal of Geophysical Research*, 98(B9), pp. 15,833–15,850. https://doi.org/10.1029/93jb01093

Horrebow, N. (1758) *The Natural History of Iceland*. Vienna: Linde.

Huerta, P., Rodríguez-Berriguete, Á., Martín-García, R. et al. (2015) The role of climate and aeolian dust input in calcrete formation in volcanic islands (Lanzarote and Fuerteventura, Spain). *Palaeogeography, Palaeoclimatology, Palaeoecology*, 417, pp. 66–79. https://doi.org/10.1016/j.palaeo.2014.10.008

Karlsson, G. (2000) *The History of Iceland*. Minneapolis: University of Minnesota Press.

Karlsson, G. (2000) *Iceland's 1100 Years: The History of a Marginal Society*. London: Hurst & Company. https://doi.org/10.1080/03468750310002272

Lambers, H., Clements, J.C. and Nelson, M.N. (2013) How a phosphorus-acquisition strategy based on carboxylate exudation powers the success and agronomic potential of lupines (Lupinus, Fabaceae). *American Journal of Botany*, 100(2), pp. 263–288. https://doi.org/10.3732/ajb.1200474

Larsen, G. and Thórarinsson, S. (1977) H4 and other acid Hekla tephra layers. *Jökull*, 27, pp. 28–46.

McGovern, T.H., Vésteinsson, O., Friðriksson, A. et al. (2007) Landscapes of settlement in northern Iceland: Historical ecology of human impact and climate fluctuation on the millennial scale. *American Anthropologist*, 109(1), pp. 27–51. https://doi.org/10.1525/aa.2007.109.1.27

Menendez, I., Derbyshire, E., Engelbrecht, J. et al. (2009) Saharan Dust and the Aerosols on the Canary Islands: Past and Present. In Cheng M and Liu W. (eds) *Airborne Particulates*, pp. 39–80. Hauppauge NY: Nova Publishers.

Middleton, N.J. (2017) Desert dust hazards: A global review, *Aeolian Research*, 24, pp. 53–63. https://doi.org/10.1016/j.aeolia.2016.12.001

Ministry for the Environment (2001) *Biological Diversity in Iceland: National Report to the Convention on Biological Diversity*. Reykjavik: Ministry for the Environment / The Icelandic Institute of Natural History.

Muschitiello, F., Pausata, F.S., Lea, J.M. et al. (2017) Enhanced ice sheet melting driven by volcanic eruptions during the last deglaciation. *Nature Communications*, 8(1), pp. 1–9. https://doi.org/10.1038/s41467-017-01273-1

Nicol, J. (1841) *An Historical and Descriptive Account of Iceland, Greenland, and the Faroe Islands*. Edinburgh: Oliver & Boyd.

Preusser, H. (1976) *The Landscapes of Iceland: Types and Regions*. The Hague: Springer Netherlands.

Runolfsson, S. (1978) Soil Conservation in Iceland. In: Holdgate, M.W. and Woodman, M.J. (eds) *The Breakdown and Restoration of Ecosystems*, pp. 231–240. NATO Conference Series, Vol 3. Springer: Boston, MA. https://doi.org/10.1007/978-1-4613-4012-6_19

Thor, J. (2002) *Icelanders in North America: The First Settlers*. Winnipeg: University of Manitoba Press.
Thorhallsdottir, T.E. (1997) Tundra ecosystems of Iceland. In Wiegolaski, F.E. (ed.) *Ecosystems of the World*. Amsterdam: Elsevier, pp. 85–96.
Thórhallsdóttir, T.E. (2002) Evaluating nature and wilderness in Iceland. In Watson, A.E., Alessa, L. and Sproull, J. *Wilderness in the Circumpolar North: Searching for Compatibility in Ecological, Traditional, and Ecotourism Values*, pp. 96–104. United States Department of Agriculture, Forest Service, Rocky Mountain Research Station.
Traustason, B. and Snorrason, A. (2008) Spatial distribution of forests and woodlands in Iceland in accordance with the CORINE land cover classification. *Icelandic Agricultural Sciences* 21, pp. 39–47.
Vergadi, E., Rouva, G., Angeli, M. et al. (2022) Infectious diseases associated with desert dust outbreaks: A systematic review. *International Journal of Environmental Research and Public Health*, 19(11), p. 6907. https://doi.org/10.3390/ijerph19116907
Vetter, V.M.S., Tjaden, N.B., Jaeschke, A. et al. (2018) Invasion of a legume ecosystem engineer in a cold biome alters plant biodiversity. *Frontiers in Plant Science*, 9, p. 715. https://doi.org/10.3389/fpls.2018.00715
World Health Organization (2021) *WHO Global Air Quality Guidelines*. Executive summary. Geneva: World Health Organization.

7. United Kingdom: Breckland – An Arid Desert in Green East Anglia

Allison, K.J. (1957) The sheep-corn husbandry of Norfolk in the sixteenth and seventeenth centuries. *The Agricultural History Review*, 5(1), pp. 12–30.
Alonso, I. (2015) Where are we now and what is the gap? In Alonso, I., Underhill-Day, J. & Lake, S. (eds) *Proceedings of the 11th National Heathland Conference, 18–20 March 2015. Sunningdale Park, Berkshire*, pp. 79–86.
Antoine, P., Catt, J., Lautridou, J.P. et al. (2003) The loess and coversands of northern France and southern England. *Journal of Quaternary Science*, 18(3–4), pp. 309–318. https://doi.org/10.1002/jqs.750
Bailey, M. (2008) *A Marginal Economy? East Anglian Breckland in the Later Middle Ages*. Cambridge: Cambridge University Press. https://doi.org/10.1017/CBO9780511896477
Bamford, E. (1996) *Tiptree: One Day a City? The Story of Tiptree Part One*. Tollesbury, Essex: pubd. Elaine Bamford/Owl Printing.
Bell, D., Endean, J. & Mountjoy, P. (2021) *Techniques to encourage European rabbit recovery*. Norwich: Back from the Brink, Natural England and University of East Anglia.
Bewick, T. (1862) *A Memoir of Thomas Bewick*. London: Longman, Green, Longman, and Roberts.
Bertran, P., Bosq, M., Borderie, Q. et al. (2021) Revised map of European aeolian deposits derived from soil texture data. *Quaternary Science Reviews*, 266, pp. 107085–107085. https://doi.org/10.1016/j.quascirev.2021.107085
Boot, F. and Davenport, A. (1977) *The Creation of a Village: The Story of Tiptree*. Tiptree: WEA.
Braunmuller, A.R. (ed.) (1997) *Macbeth: The New Cambridge Shakespeare*. Cambridge: Cambridge University Press.
Breckland Society (2007) *The Vernacular Architecture of Breckland: A Survey by The Breckland Society*. Pub. The Breckland Society, King's Lynn.
Breckland Society (2016) *The Military History of the Brecks 1900–1949: A Report by The Breckland Society*. King's Lynn: The Breckland Society.
Cairns, V., Wallenhorst, C., Rietbrock, S. et al. (2019) Incidence of Lyme disease in the UK: A population-based cohort study. *BMJ Open*, 9(7), e025916. https://doi.org/10.1136/bmjopen-2018-025916
Castle, D.A., McCunnall, J., and Tring, I.M. (1984) *Field Drainage: Principles and Practices*. Batsford, London.
Chatters, C. (2021) *Heathland*. Bloomsbury Publishing.
Chippaux, J.-P. (2012) Epidemiology of snakebites in Europe: A systematic review of the literature. *Toxicon*, 59(1), pp. 86–99. https://doi.org/10.1016/j.toxicon.2011.10.008
Clare, J., Blunden, E. and Porter, A. (1920) *John Clare Poems*. London: Richard Comden-Sanderson.
Clare, J. (1975) *Selected Poems and Prose of John Clare*. Oxford: Clarendon.
Clare, J. and Symons, A. (1908) *Poems by John Clare*. London: Henry Frowde.
Clare, J., Blunden, E. and Porter, A. (1920) *John Clare Poems*. London: Richard Comden-Sanderson.
Clare, J.B. (1903) *Curious Parish Records and Workhouse Riots*. Halesworth: William P. Gale.
Clark, M. (2011) The gentry, the commons, and the politics of common right in Enfield c.1558–c.1603. *The Historical Journal*, 54, pp. 609–629. https://doi.org/10.1017/S0018246X11000185

Clarke, W.G. (1914) The Breckland sand-pall and its vegetation. *Transactions of Norfolk and Norwich Naturalists Society*, X(2), p. 130.
Cobbett, W. (1893) *Rural Rides Vol. 1*. London: Reeves and Turner.
Cook, O. (1980) *Breckland*. London: Robert Hale.
Couch, C. (2021) Other expanded towns serving London. In Crouch, C. *Planned Urban Development*, pp. 98–126. Cheltenham: Edward Elgar Publishing.
Crow, P. (2004) *Trees and Forestry on Archaeological Sites in the UK: A Review Document*. Alice Holt: Forest Research.
Darby, H.C. (ed.) (1938) *The Cambridge Region*. Cambridge: Cambridge University Press.
Davison, A. (1996) *Deserted Villages in Norfolk*. North Walsham: Poppyland.
De Keyzer, M. and Bateman, M.D. (2018) Late Holocene landscape instability in the Breckland (England) drift sands. *Geomorphology*, 323, pp. 123–134.
de la Rochefoucauld, F. (1933) *A Frenchman in England, 1784*. Cambridge: Cambridge University Press.
Dickson, N. and The Brecks Partnership (2013) *Breaking New Ground: A Landscape Conservation Action Plan for The Brecks*. Online PDF at http://www.breakingnewground.org.uk/assets/LCAP/BNGLPS-Landscape-Conservation-Action-Plan-web.pdf
Dittner, L. and Parry, J. (eds) (2017) Editorial note on definitions and nomenclature. *Journal of Breckland Studies*, 1, p. 5.
Dolman, P. and Sutherland, W. (1991) Historical clues to conservation. *New Scientist*, 751, pp. 40–43.
Dolman, P.M. and Sutherland, W.J. (1992) The ecological changes of Breckland grass heaths and the consequences of management. *Journal of Applied Ecology*, 29(2), pp. 402–413.
Dolman, P.M., Panter, C.J., and Mossman, H.L. (2010) *Securing Biodiversity in Breckland: Guidance for Conservation and Research. First Report of the Breckland Biodiversity Audit*. University of East Anglia, Norwich.
Durkheim, E. (1995) *The Elementary Forms of Religious Life*. New York: Free Press.
Dyndor, Z. (2015) The gibbet in the landscape: Locating the criminal corpse in mid-eighteenth-century England. In Ward, R. (ed.) *A Global History of Execution and the Criminal Corpse*, pp. 102–125. London: Palgrave Macmillan.
Environment Agency/Natural England (2015) Improvement Programme for England's Natura 2000 Sites (IPENS) Planning for the Future: Site Improvement Plan – Breckland. At https://publications.naturalengland.org.uk/publication/5075188492271616?category=4873023563759616 / http://www.naturalengland.org.uk/ipens2000
Essex Wildlife Trust (2022) *Tiptree Heath*. Colchester: Essex Wildlife Trust.
Evelyn, J. (1901) *The Diary of John Evelyn: Vol. II*. Washington: M. Walter Dunne.
Farrell, L. (1993) *Lowland Heathland: The Extent of Habitat Change*. Peterborough: English Nature.
Field, J. (2009) Able bodies: Work camps and the training of the unemployed in Britain before 1939. Paper given at conference *The Significance of the Historical Perspective in Adult Education Research*, University of Cambridge Institute of Continuing Education, 6 July 2009.
Field, J. (2012) An Anti-Urban Education? Work camps and ideals of the land in Interwar Britain. *Rural History*, 23(2), pp. 213–228. https://doi.org/10.1017/s0956793312000088
Field, J. (2014) Incremental growth: Instructional Centres under the National Government'. In Field, J. *Working Men's Bodies: Work Camps in Britain, 1880–1940*. Manchester: Manchester University Press. https://doi.org/10.7228/manchester/9780719087684.003.0008
Fischer, N. and Küster, H. (2023) The NW German heathland: A threatened landscape? *Revista Murciana de Antropología*, 30, pp. 15–36.
Fletcher, D.R. (2013) The governance of social marginality in the UK: Towards the centaur state? *British Journal of Community Justice*, 11(1), pp. 19–34.
Foucault, M., 1984. Of other spaces: Utopias and Heterotopias. *Architecture/Mouvement/Continuité*, 5(1), pp. 1–9.
Gardner, E., Julian, A., Monk, C. et al. (2019) Make the adder count: Population trends from a citizen science survey of UK adders. *Herpetological Journal*, 29, pp. 57–70. https://doi.org/10.33256/hj29.1.5770
Gilpin, W. (1802) *An Essay on Prints*. London: Printed for Cadell & Davies.
Gilpin, W. (1809) *Observations on Several Parts of the Counties of Cambridge, Norfolk, Suffolk, and Essex*. London: Printed for Cadell and Davies.
Gray, J., Kahl, O. and Zintl, A. (2021) What do we still need to know about Ixodes ricinus? *Ticks and Tick-Borne Diseases*, 12(3), p. 101682. https://doi.org/10.1016/j.ttbdis.2021.101682
Griffin, C.J. (2023) Enclosure as internal colonisation: The subaltern commoner, terra nullius and the settling of England's 'wastes'. *Transactions of the Royal Historical Society*, 1, pp. 95–120. https://doi.org/10.1017/S0080440123000014

Hawkes, R.W., Smart, J., Brown, A. et al. (2021) Effects of experimental land management on habitat use by Eurasian Stone-curlews. *Animal Conservation*, 24(5), pp. 743–755. https://doi.org/10.1111/acv.12678

HM Government (2023) *Environmental Improvement Plan 2023: First revision of the 25 Year Environment Plan*. London: Department for Environment, Food and Rural Affairs.

Hoskins, W.G. (1955) *The Making of the English Landscape*. London: Hodder and Stoughton.

Hoskins W.G. (1978) *One Man's England*. London: BBC.

Hoskins. W.G. and Dudley Stamp, L. (1963) *The Common Lands of England and Wales*. London: Collins.

Howkins, C. (1997) *Heathland Harvest*. Addlestone, Surrey: Pub. Chris Howkins.

Humphries J. (1990) Enclosures, Common Rights, and Women: The proletarianization of families in the late eighteenth and early nineteenth centuries. *The Journal of Economic History*, 50(1), pp. 17–42. https://doi.org/10.1017/s0022050700035701

Instructional Centres Hansard HC Deb vol. 346 cc2036–8, 4 May 1939.

Jackson, R.H. and Henrie, R. (1983) Perception of sacred space. *Journal of Cultural Geography*, 3(2), pp. 94–107. https://doi.org/10.1080/08873638309478598

Jessup, G. (1992) *Breckland Ramblings*. Wymondham: Geo. R. Reeve.

Johnceline, K. (1998) *Wenhaston: Millennial History of a Suffolk Village*. Wenhaston: Wyvern Publishing.

Kahl, O. and Gray, J.S. (2022) The biology of *Ixodes ricinus* with emphasis on its ecology. *Ticks and Tick-Borne Diseases*, 14(2), p. 102114. https://doi.org/10.1016/j.ttbdis.2022.102114

Kavanagh, G. (ed.) (2001) *Making City Histories in Museums*. Leicester: Leicester University Press.

King, D. (1999) Reconditioning the unemployed: Work camps in Britain. In King, Desmond *In the Name of Liberalism: Illiberal Social Policy in the USA and Britain*, pp. 155–179. Oxford: Oxford Academic. https://doi.org/10.1093/0198296290.003.0008

Kirschner, P., Záveská, E., Gamisch, A. et al. (2020) Long-term isolation of European steppe outposts boosts the biome's conservation value. *Nature Communications*, 11(1). https://doi.org/10.1038/s41467-020-15620-2

Land Use Consultants (2007) *Landscape Character Assessment of Breckland District: Final Report*. Prepared for Breckland Council by Land Use Consultants, London.

Lewis, C. (2016) Disaster recovery: New archaeological evidence for the long-term impact of the 'calamitous' fourteenth century. *Antiquity*, 90(351), pp. 777–797. https://doi.org/10.15184/aqy.2016.69

Martin, J. (2010) The wild rabbit: plague, polices and pestilence in England and Wales, 1931–1955. *Agricultural History Review*, 58(2), pp. 255–276.

Mason, A., Parry, J. and Dittner, L. (2010) *The Warrens of Breckland: A Survey by The Breckland Society*. King's Lynn: The Breckland Society/English Heritage.

Mason, A., Parry, J. and Dittner, L. (eds) (2016) *Flint in The Brecks*. Thetford: The Breckland Society.

Matless, D. (2008) Properties of ancient landscape: The present prehistoric in twentieth-century Breckland. *Journal of Historical Geography*, 34(1), pp. 68–93.

Mendel, H. and Parsons, E. (2016) Observations on the life-history of the Silver-studded Blue, Plebejus argus L. *Trans. Suffolk Nat. Soc.*, 23, pp. 2–8.

Mitchell, J. (2017) Father left fighting for life after being bitten by adder in west London park. *The Standard* newspaper (London), 12 April 2017.

Moody D. (2016) Godshillwood and Woodgreen: A squatter settlement on the edge of the New Forest 1600–1840. *Proc. Hampshire Field Club Archaeol. Soc.*, 71, pp. 126–147.

Moseley, M.J. (1973) Some problems of small expanding towns. *The Town Planning Review*, 44(3), pp. 263–278.

Natural England (2015) *National Character Area profile: 85. The Brecks*. Worcester: Natural England.

Newton, A.C., Stewart, G.B., Myers, G. et al. (2009) Impacts of grazing on lowland heathland in north-west Europe. *Biological Conservation*, 142(5), pp. 935–947. https://doi.org/10.1016/j.biocon.2008.10.018

Panter, C.J., Mossman, H.L., and Dolman, P.M. (2013) *Stanford Training Area (STANTA) Biodiversity Audit to Support Grass Heath Management*. Norwich: University of East Anglia.

Powell, D. (2009) *First Publications of John Clare's Poems*. Research Papers on John Clare, number 1. Missoula: The John Clare Society of North America, 2nd edition.

Randall, R. and Dymond, D. (1996) Why Thetford Forest? The human and natural history of Breckland before the early 20th century. In Ratcliffe, P. and Claridge, J. (eds) *Thetford Forest Park: The Ecology of a Pine Forest*, pp. 1–5. Forestry Commission Technical Paper 13, Forestry Commission, Edinburgh.

Ratcliffe, P. and Claridge, J. (eds) (1996) *Thetford Forest Park: The Ecology of a Pine Forest*. Forestry Commission Technical Paper 13. Edinburgh: Forestry Commission.

Schulte, R.P.O., Fealy, R., Creamer, R.E. et al. (2012) A review of the role of excess soil moisture conditions in constraining farm practices under Atlantic conditions. *Soil Use and Management*, 28(4), pp. 580–589.

Sheail, J. (1971) *Rabbits and their History*. Exeter: David & Charles.

Sheail, J. and Bailey, M. (1996) The History of the Rabbit in Breckland. In Ratcliffe, P. and Claridge, J. (eds) *Thetford Forest Park: the Ecology of a Pine Forest*. Forestry Commission Technical Paper 13, Forestry Commission, Edinburgh.
Shearwood, M. (2022) A summer on Hounslow Heath: Combined army operations in late-seventeenth-century England. *War in History*, 29(3), pp. 525–542. https://doi.org/10.1177/09683445211029974
Shielsflynn (undat.) *Brecks' Special Qualities: An Analysis of Identity and Sense of Place*. Shielsflynn, Cambridge. Online at http://www.sheilsflynn.com
Shielsflynn (undat.) *Brecks' Special Qualities: An analysis of identity and sense of place*. Cambridge: Shielsflynn Studio, at http://www.breakingnewground.org.uk/assets/LCAP/Brecks-Special-Qualities-Report-low-res.pdf
Skipper, K. and Williamson, T. (1997) *Thetford Forest: Making A Landscape 1922–1997*. Norwich: Centre of East Anglia Studies, University of East Anglia.
St. John, J. (1787) *Observations on the Land Revenues of the Crown*. Appx. 3. London: Printed for J. Debrett.
State of Nature Partnership (2023) *State of Nature 2023 Report*. TP25999. United Kingdom: State of Nature Partnership. https://doi.org/10.500.12592/kc8x73
Tarlow, S. (2014) The Technology of the Gibbet. *International Journal of Historical Archaeology*, 18, pp. 668–699. https://doi.org/10.1007/s10761-014-0275-0
Tarlow, S. (2017) *The Golden and Ghoulish Age of the Gibbet in Britain*. London: Palgrave Macmillan. https://doi.org/10.1057/978-1-137-60089-9
Tarlow, S. and Dyndor, Z. (2015) The landscape of the gibbet, *Landscape History*, 36(1), pp. 71–88. https://doi.org/10.1080/01433768.2015.1044284
Tolley, D. (2017) Holst, Hardy, the heath, and an eclipse. *The Hardy Society Journal*, 13(1), pp. 72–75.
Török, P., Neuffer, B., Heilmeier, H. et al. (2020) Climate, landscape history and management drive Eurasian steppe biodiversity. *Flora*, 271, p. 151685. https://doi.org/10.1016/j.flora.2020.151685
Townshend, D., Stace, H., and Radley, D. (2004) *State of Nature: Lowlands – Future Landscapes for Wildlife*. Peterborough: English Nature.
Waites, I. (2006) Darkness terrible in its own nature: Turner's Sublime in the common heathlands of South East London c.1796–7. In: *Romantic Spectacle Conference*. Centre for Research in Romanticism, 7–9 July 2006, Roehampton University, London.
Waites, I. (2012) *Common Land in English Painting 1700–1850*. Woodbridge: Boydell Press. https://doi.org/10.1017/S0956793313000265
Ward, C. (2002) *Cotters and Squatters: Housing's Hidden History*. Nottingham: Five Leaves.
Wesche, K., Ambarlı, D., Kamp, J. et al. (2016) The Palaearctic steppe biome: A new synthesis. *Biodiversity and Conservation*, 25(12), pp. 2197–2231. https://doi.org/10.1007/s10531-016-1214-7
Whyte, N. and Hoyle, R.W. (2011) Contested Pasts: Custom, Conflict and Landscape Change in West Norfolk, c.1550–1650. In: Hoyle, R.W. (ed.) *Custom, Improvement and the Landscape in Early Modern Britain*, pp. 101–26. London: Routledge.
Williamson, T. (2007) *Rabbits, Warrens and Archaeology*. Stroud: Tempus.
Wright, T. (1668) A curious and exact relation of a sand-floud, which hath lately overwhelmed a great tract of land in the county Suffolk; together with an account of the check in part given to it; communicated in an obliging letter to the Publisher, by that Worthy Gentleman Thomas Wright Esquire, living upon the place, and a sufferer by that Deluge. *Philosophical Transactions of the Royal Society of London*, 3(37), pp. 722–725. https://doi.org/10.1098/rstl.1668.0025

8. Germany: A Sandy Hike Through Purple Heather

Aerts, R. and Heil, G.W. (eds) (1993) *Heathlands: Patterns and Processes in a Changing Environment*. Dordrecht: Kluwer Academic Publishers.
Alonso, I., Underhill-Day, J. and Lake, S. (eds) (2015) *Proceedings of the 11th National Heathland Conference, 18–20 March 2015*. Sunningdale Park, Berkshire.
Anon. (2009) The use of prescribed fire for maintaining open *Calluna* heathlands in North Rhine Westphalia, Germany. *International Forest Fire News*, no. 38, January–December 2009, pp. 75–80.
Ascoli, D., Beghin, R., Ceccato, R. et al. (2009) Developing an adaptive management approach to prescribed burning: A long-term heathland conservation experiment in north-west Italy. *International Journal of Wildland Fire*, 18(6), p. 727–735. https://doi.org/10.1071/wf07114
Baggesen, J. (1986) *Das Labyrinth oder Reise durch Deutschland in die Schweiz 1789*. Munich: Beck.
Cucu A-A, Baci G-M, Cucu A-B. et al. (2022) *Calluna vulgaris* as a valuable source of bioactive compounds: Exploring Its phytochemical profile, biological activities and apitherapeutic potential. *Plants*, 11(15), p. 1993. https://doi.org/10.3390/plants11151993

Fischer, N. (ed.) (2023) The NW German heathland: A threatened landscape? *Revista Murciana de Antropología*, 30, pp. 15–36. https://doi.org/10.6018/rmu.512661

Gimingham, C.H. (1972) *Ecology of Heathlands*. London: Chapman and Hall.

Gimingham, C.H. (1975) *An Introduction to Heathland Ecology*. Edinburgh: Oliver and Boyd.

Green, B.H. (1972) The relevance of seral eutrophication and plant competition to the management of successional communities. *Biological Conservation*, 4(5), pp. 378–384. https://doi.org/10.1016/0006-3207(72)90057-2

Hochkirch, A., Gärtner, A.C. and Brandt, T. (2008) Effects of forest-dune ecotone management on the endangered heath grasshopper, *Chorthippus vagans* (Orthoptera: Acrididae). *Bulletin of Entomological Research*, 98(5), pp. 449–456. https://doi.org/10.1017/s0007485308005762

Joint Nature Conservation Committee (2009) *Common Standards Monitoring Guidance for Lowland Heathland*. Version February 2009. Peterborough: Joint Nature Conservation Committee.

Keienburg, T. and Prüter, J. (2004) Project: The management of heathlands in northwest Germany (Luneberger Heide nature reserve) by prescribed burning in winter. *European Fire in Nature Conservation Network*. https://gfmc.online/wp-content/uploads/NNA-Project-new-Sep-2004-1.pdf

Keienburg, T., & Prüter, J. (2006) *Safeguarding the heathlands of Europe – Lüneburg Heath Nature Reserve: Preservation and Development of an Old Cultural Landscape*. Seim: The Heathland Centre.

Koster, E.A. (2009) The "European Aeolian Sand Belt": Geoconservation of Drift Sand Landscapes. *Geoheritage*, 1(2), pp. 93–110. https://doi.org/10.1007/s12371-009-0007-8

Leuschner, C., and Ellenberg, H. (2017) The Central European Vegetation as the Result of Millennia of Human Activity. In: Leuschner, C., and Ellenberg, H. *Ecology of Central European Forests*, pp. 31–116. Cham: Springer. https://doi.org/10.1007/978-3-319-43042-3_3

Mangourit, M.O.B. (1806) *Travels in Hanover, During the Years 1803 and 1804*. London. Printed for Richard Phillips.

Mathias, E., Czerkus, G. and Schenk, A. (2022) *The Role of Pastoralism in Germany*. Ober-Ramstadt: League for Pastoral Peoples and Endogenous Livestock Development.

Natural England (1996) *The Lowland Heathland Management Booklet Version 2*. ENS11. Sheffield; Natural England.

Reim, S., Lochschmidt, F., Proft, A. et al. (2016) Genetic structure and diversity in *Juniperus communis* populations in Saxony, Germany. *Biodiversity Research and Conservation*, 42(1), pp. 9–18. https://doi.org/10.1515/biorc-2016-0008

Seibert, B., Gauly, M., and Erhardt, G. (2004) Productivity of different sheep breeds in extensive pasture management. *Archiv fur Tierzucht*, 47(6), pp. 142–152.

Shellswell, C.H., Chant J.J., Alonso, I. et al. (2016) *Restoration of Existing Lowland Heathland: Timescales to Achieve Favourable Condition*. Salisbury: Plantlife.

Stafford, R., Chamberlain, B., Clavey, L. et al. (eds) (2021) *Nature-based Solutions for Climate Change in the UK. Technical Report*. London: British Ecological Society.

Thomas, P.A., El-Barghathi, M. and Polwart, A. (2007) Biological flora of the British Isles: *Juniperus communis* L. *Journal of Ecology*, 95(6), pp. 1404–1440. https://doi.org/10.1111/j.1365-2745.2007.01308.x

Urban, B., Kunz, A. and Gehrt, E. (2011) Genesis and dating of Late Pleistocene-Holocene soil sediment sequences from the Lüneburg Heath, Northern Germany. *Quaternary Science Journal*, 60(1), pp. 6–26. https://doi.org/10.3285/eg.60.1.01

Watt, M.S., Clinton, P.W., Whitehead, D. et al. (2003) Above-ground biomass accumulation and nitrogen fixation of broom (*Cytisus scoparius* L.) growing with juvenile *Pinus radiata* on a dryland site. *Forest Ecology and Management*, 184(1–3), pp. 93–104. https://doi.org/10.1016/s0378-1127(03)00151-8

Webb, N.R. (1998) History and ecology of European heathlands. *Transactions Suffolk Nat. Soc.*, 34, p. 1–8.

Woestenburg, M. (2018) Heathland farm as a new commons? *Landscape Research*, 43(8), pp. 1045–1055. https://doi.org/10.1080/01426397.2018.1503236

Conclusion – European Deserts: Perspectives, and a Vision of the Future

Brakensiek, S. (2002) The management of common land in north western Germany. In De Moor, M., Shaw-Taylor, L., and Warde, P. (eds) *The Management of Common Land in north west Europe, c.1500–1850*, pp. 225–245. Turnhout, Belgium: Brepols Publishers.

Chatters, C. (2021) *Heathland*. Bloomsbury Publishing.

Crofts, R. (2000) Scotland's Cultural Landscape. Paper to the *Cultural Landscape: Planning for a Sustainable Partnership Between People and Place* conference, 3–5 May, 2000, Rewley House, Oxford. ICOMOS: Oxford.

Diemont, W.H., Webb, N. and Degn, H.J. (1996) A pan European view on heathland conservation. In Anon. *Proceedings of the National Heathland Conference, 18–20 September 1996, Hampshire, UK*, pp. 21–32.

Dragotă, C.S., Dumitrașcu, M., Kucsica, G. et al. (2011) Assessing dryness and drought phenomena in the south Oltenia. In *Proceedings of the 12th International Conference on Environmental Science and Technology*, Rhodos, Grecia, pp. A448–455.

Dragotă, C.S., Dumitrașcu, M., Grigorescu, I. et al. (2011) The climatic water deficit in South Oltenia using the Thornthwaite method. *Forum Geografic*, 10(1), pp. 140–148. https://doi.org/10.5775/fg.2067-4635.2011.032.i

Duddigan, S., Hales-Henao, A., Bruce, M. et al. (2024) Restored lowland heathlands store substantially less carbon than undisturbed lowland heath. *Communications Earth & Environment*, 5(15). https://doi.org/10.1038/s43247-023-01176-8

Dumitrașcu, M., Grigorescu, I., Cuculici, R. et al. (2014) Assessing long-term changes in forest cover in the South West Development Region, Romania. *Forum Geografic*, 13(1), pp. 76–85. https://doi.org/10.5775/fg.2067-4635.2014.159.i

Dusca, A. and Badea, S. (2009) The causes of the Romanian desert and means of avoiding a possible human migration. In Anon. *IOP Conference Series. Earth and Environmental Science*, 6(56/562011). Bristol: IOP Publishing. https://doi.org/10.1088/1755-1307/6/56/562011

Ellis, E.C. (2011) Anthropogenic transformation of the terrestrial biosphere. *Philosophical Transactions of the Royal Society A: Mathematical, Physical and Engineering Sciences*, 369(1938), pp. 1010–1035. https://doi.org/10.1098/rsta.2010.0331

Ellis, E.C. (2013) Sustaining biodiversity and people in the world's anthropogenic biomes. *Current Opinion in Environmental Sustainability*, 5(3–4), pp. 368–372. https://doi.org/10.1016/j.cosust.2013.07.002

Eriksson, O. (2021) The importance of traditional agricultural landscapes for preventing species extinctions. *Biodiversity and Conservation*, 30(5), pp. 1341–1357. https://doi.org/10.1007/s10531-021-02145-3

European Environment Agency (2019) *The European Environment – State and Outlook 2020: Knowledge for Transition to a Sustainable Europe*. Luxembourg: Publications Office of the European Union.

European Environment Agency (2020) *State of Nature in the EU: Results from reporting under the nature directives 2013–2018. EEA Report No 10/2020*. Luxembourg: Publications Office of the European Union.

Ezcurra, E. (ed.) (2006) *Global Deserts Outlook*. Nairobi: UNEP.

Fagúndez, J. (2012) Heathlands Confronting global change: Drivers of biodiversity loss from past to future scenarios. *Annals of Botany*, 111(2), pp. 151–172. https://doi.org/10.1093/aob/mcs257

Gimingham, C.H. (1992) *The Lowland Heathland Management Book part 1*. Peterborough: English Nature Science.

Gregg, R., Elias, J.L., Alonso, I. et al. (2021) *Carbon storage and Sequestration by Habitat: A Review of the Evidence*. Second edition. Natural England Research Report NERR094. York: Natural England.

Griffin, C.J. (2023) Enclosure as internal colonisation: The subaltern commoner, terra nullius and the settling of England's 'wastes'. *Transactions of the Royal Historical Society*, 1, pp. 95–120. https://doi.org/10.1017/S0080440123000014

Haveman, R., van der Berg, A. and van der Wijngaart, R. (2005) Abandonment of sod-cutting may cause loss of characteristic heathland communities – The case of a military training area in the Netherlands. *Annali Di Botanica*, 5, pp. 161–170. https://doi.org/10.4462/annbotrm-9216

Hawley, G., Anderson, P., Gash, M. et al. (2008) *Impact of Heathland Restoration and Re-Creation Techniques on Soil Characteristics and the Historical Environment*. Natural England Research Reports, Number 010. Number 010. York: Natural England.

Hopkins, J. (2009) Climate change adaptation of heathland biodiversity. In Alonso, I. *Proceedings of the 10th National Heathland Conference – Managing Heathlands in the Face of Climate Change, 9–11 September 2008, University of York*. Natural England Commissioned Report NECR014. pp. 2–14. Sheffield: Natural England.

Johnson, A., Hebdon, C., Burow, P. et al. (2022) Anthropocene. In *Oxford Research Encyclopedia of Anthropology*. https://doi.org/10.1093/acrefore/9780190854584.013.295

Joint Nature Conservation Committee (2009) *Common Standards Monitoring Guidance for Lowland Heathland*. Version February 2009. Peterborough: Joint Nature Conservation Committee.

Keienburg, T., & Prüter, J. (2006) *Safeguarding the heathlands of Europe – Lüneburg Heath Nature Reserve: Preservation and Development of an Old Cultural Landscape*. Seim: The Heathland Centre.

Krings, W. (1986) Rural Colonization in Western Central Europe from the End of the 19th Century up to the Present: Ambitious Plans – Disappointing Results? *Erdkunde Bd. 40*, H. 3, pp. 227–235.

BIBLIOGRAPHY

Loidi, J., de Blust, G., Campos, J.A. et al. (2019) Heathlands of Temperate and Boreal Europe. In Elias, S.A. (ed.) *Reference Module in Earth Systems and Environmental Sciences*. Amsterdam: Elsevier. https://doi.org/10.1016/B978-0-12-409548-9.12078-0

Meek, M.H., Beever, E.A., Barbosa, S. et al. (2022) Understanding Local Adaptation to Prepare Populations for Climate Change. *BioScience*, 73(1), pp. 36–47. https://doi.org/10.1093/biosci/biac101

Mitrică, B., Grigorescu, I., Mocanu, I. et al. (2012) A comparative approach for assessing water-related environmental issues in Oltenia and Banat Plains, Romania. In Anon. *Proceedings of the BALWOIS 2012 Fifth International Scientific Conference on Water, Climate and Environment*. Balkan Institute for Water and Environment, Ohrid, Republic of Macedonia, 27 May–2 June 2012.

Musset, R. (1939) La production du fumier de ferme en France. *Annales de Geographie*, 48(272), pp. 202–203.

Natural England and RSPB (2019) *Climate Change Adaptation Manual – Evidence to Support Nature Conservation in a Changing Climate*. NE751. Second Edition. Natural England, York, UK.

Olmeda, C., Šefferová, V., Underwood, E., Millan, L., Gil, T. and Naumann, S. (eds) (2020) *EU Action Plan to Maintain and Restore to Favourable Conservation Status the Habitat Type 4030 European Dry Heaths*. Luxembourg: European Commission, Directorate-General Environment.

Paterson, B. (2006) Ethics for wildlife conservation: Overcoming the human–nature dualism. *Bioscience*, 56(2), pp. 144–150. https://doi.org/10.1641/0006-3568(2006)056[0144:efwcot]2.0.co;2

Pirret, J.S.R., Fung, F., Lowe, J.A. et al. (2020). *UKCP Factsheet: Soil Moisture*. Exeter: Met Office.

Raunkiaer, S. (1934) *The Life Forms of Plants and Statistical Plant Geography*. Oxford: Clarendon Press.

Spinoni, J., Barbosa, P., Dosio, A. et al. (2018) Is Europe at risk of desertification due to climate change? In *EGU General Assembly Conference Proceedings: The 20th EGU General Assembly, EGU2018*, 4–13 April 2018, Vienna, p. 9557 (abstract).

Staddon, P.L., Thompson, P & Short, C. (2023) *Re-evaluating the Sensitivity of Habitats to Climate Change*. NECR478. Natural England.

Stanila, A.L., Simota, C.C. and Dumitru, M.I. (2020) Contributions to the knowledge of sandy soils from Oltenia Plain. *Revista de Chimie*, 71, pp. 192–200. https://doi.org/10.37358/RC.20.1.7831

Vicente-Serrano, S.M., Pricope, S.M., Toreti, A. et al. (2024) *The Global Threat of Drying Lands: Regional and Global Aridity Trends and Future Projections*. Bonn: United Nations Convention to Combat Desertification (UNCCD). Bonn, Germany.

Waters, C.N., and Turner, S.D. (2022) Defining the onset of the Anthropocene. *Science*, 378(6621), pp. 706–708. https://doi.org/10.1126/science.ade2310

Webb. N. (1998) History and ecology of European heathlands. In *Heathland: A Waste of Space? Transactions of Suffolk Naturalists' Society*, 34, p. 1–8. https://doi.org/10.1111/j.1365-2664.1998.tb00020.x

Willner, W., Moser, D., Plenk, K. et al. (2021) Long-term continuity of steppe grasslands in eastern Central Europe: Evidence from species distribution patterns and chloroplast haplotypes. *Journal of Biogeography*, 48(12), pp. 3104–3117. https://doi.org/10.1111/jbi.14269

Woodbridge, J., Fyfe, R.M., Roberts, N. et al. (2014) The impact of the Neolithic agricultural transition in Britain: A comparison of pollen-based land-cover and archaeological 14C date-inferred population change. *Journal of Archaeological Science*, 51, pp. 216–224. https://doi.org/10.1016/j.jas.2012.10.025

Woronko, B., Dąbski, M. (2022). The North European Plain. In: Oliva, M., Nývlt, D., Fernández-Fernández, J.M. (eds) *Periglacial Landscapes of Europe*. Cham: Springer. https://doi.org/10.1007/978-3-031-14895-8

Zeeberg, J. (2008) The European sand belt in eastern Europe – and comparison of Late Glacial dune orientation with GCM simulation results. *Boreas*, 27(2), pp. 127–139. https://doi.org/10.1111/j.1502-3885.1998.tb00873.x

INDEX

Adder (European viper)   181–3
aeolian processes   12, 71, 85, 88, 96, 115, 129, 130, 135, 136, 138, 139, 142, 197
afforestation   77, 83, 128–30, 173, 186
Aldrin, Buzz   143
Antarctica, desert distribution in   17–19
   hyper-arid Antarctica   20
   major life zones   21
   polar desert   20
Anthropocene, European deserts and   222–3
Apeldoorn   65
Arctic Fox   145
Arguedas village   39–40
Arid Britain   157–8
aridity   15–17
   edaphic aridity   16
   soil aridity   16
Atacama Desert   11
Atlantic Puffin   146
atmospheric nitrogen deposition   87–90

badlands   46–53
   Bardenas badlands   44
   Golden Eagle   53
   wildlife in   50
   See also Iberian badlands
Baggesen, Jens   203
Bardenas Reales region   4, 24, 39–63
   age of   56
   agriculture   54–6
   badlands   44
   Bardenas Reales Biosphere Reserve   42
   cañadas reales track   54
   Castildetierra   41–2
   ecosystem   49–50
   environment   41–6
   grazing   58
   history   57–60
   human landscape   58
   La Sanmiguelada festival   54
   military   60–1
   populating deserts   61–3
   Pyrenees mountain range   55
   Sanchicorrota   59–60
   transhumance   54

Bearberry   118, 120
bees, in German deserts   190–5
biodiversity   15, 19, 21, 28, 41, 49–50, 58, 130, 160, 163, 166, 174, 187, 188–9, 213–14, 215–16, 217, 220, 221, 223, 227
   loss of   56, 87, 88, 197, 212
biomes   3, 21
   aquatic   21
   desert   4, 8, 21
   forest   21
   grassland   21
   internal variation   21
   tundra   21
birdlife of Netherlands deserts   75–7
   Eurasian Nightjar   75–6
Birds of Norfolk, with Remarks on their Habits, Migration, and Local Distribution, The (Stevenson & Southwell)   165
Black-bellied Sandgrouse   51
Błędów desert   112–13
   afforestation   128–30
   environmental conditions   122–6
   fulgurites   127
   galena ore mining   114–15
   looming and sandstorms   126–7
   military range   127–8
      Second World War   127
   mining history   114–15
   nature reserve designation   130–1
   Perennial Knawel plant   118
   plant life   118
   species range   116–17
   temperatures   123–5
   ultraviolet-B radiative component   123
   water   124
   See also Polish Cochineal species
Book of the Icelanders   141
Breckland   5, 156–89
   Breckland Thyme   166, 167
   the 'brecks'   158–60
   'Brecks'   158–61
   conservation   188–9
   environmental changes   187
   grazing   170
   heath hazards   181–4

INDEX

land-use changes 173–4
landscapes 160–3
nature conservation importance 186
rabbit in 163–9
sand drifts and floods 169–72
sheep as principal livestock 159
in twenty-first century 184–8
Broom 206
Brown Argus 163

Castildetierra 41–2
Cavenham Heath 186
Clare, John 175–6, 179
Clarke, William George 159
climate 1, 8, 15–17, 21, 22–3, 34, 44, 122, 140, 215
climate change 6, 27, 31, 128, 140, 154, 221–7
Cobbett, William 181
cold desert phenomenon 2, 6
Constable, John 175
Corsican Pine 173

d'Haussez, Baron 105
de Courson, Lamoignon 104–5
de Luc, Jean André 203
de Malla, Gaspar 60
desert biome of the world 8–38
environmental variables 22–3
global desert-dryland biome 14
human dimension 27–32
meaning 31–2
political/administrative borders 10
population density in 10
risks in 9
significance 31–2
desert dust impacts 147–8
desert island 132, 133–41
black-sand wildernesses 134
desertic Iceland 155
deserts
Antarctica 17–19
distribution 17–22
and dry lands 11–17
United Nations definition 16
Diamond, Jared 140
Dingy Skipper 163
Dornon, M. Sylvain 102
drift sands decline 83–7
dry lands 11–17
arid 13
dry subhumid 13, 15, 122
hyper-arid 13
semi-arid 13
Dupont's Lark 51
dust 12
Dyngjusandur 133, 138

East Anglia, Arid desert in 156–89. *See also* Breckland
edaphic aridity 16, 18, 215

Egyptian Vulture 52
environmental history, Iceland 140–3
environmental variables 22–3
Esparto Grass 53
European deserts 1–2, 21, 23–6, 215–28
and Anthropocene 222–3
archaeological value 29
climate 34
climate change and 223–6
colour palette 4
ecology 34
economic identification 35
economic redundancy 28
formal terminology and land classification 35
geography 33
geology 34
heathland as 33
hydrology 34
indicative criteria for identification 33–6
lack of moisture 24
landscape 34
located on European subcontinent 33
located inland 33
located in lowland regions 33
location 24, 33
naming 35
planning and management for the future 218–21
rationale for conservation 3
relief and geomorphology 34
semi-natural habitat type 25
sociocultural 35
soils 34
species in 26
surface water sources 25
towards a working description 32–6
vision for 226–8
vulnerable 35

Fairhair, Haraldr 148
farming 9, 35, 56, 82, 96, 99–100, 155, 159–60, 168, 188, 190, 198, 200, 208, 210, 213, 218, 222
forest 13, 21, 65, 85–6, 95, 106–8, 113, 156, 173, 211, 221–2
France 91–111
shepherds on stilts 101–2
See also Les Landes
frogs 6

geodiversity 3
geology 25, 34, 41, 44–6, 49, 70, 113, 137, 143, 157, 197
geomorphology 34
Germany 190–214
big landscapes and heather 195–200
broom and juniper 206–8

ecological management and public opinion 213–14
heath sheep and bees 190–5
See also Lüneburg Heide, management and visitors
gibbeting 180
Gilpin, William 169–70
Gimingham, Charles H. 198
Gintrac, Jean-Louis 101
glaciers 135–7
Gobi Desert 9
Golden Eagle 53
Golden Jackal 193
Grant, Ulysses S. 150
Grass Snake 6, 116, 183
Grasset de Saint-Sauveur, J. 104–5
grassland 9, 21, 33, 35, 76, 93, 160–1, 166, 168, 224
Grayling butterfly 162–3
Great Bustard 26, 53, 160–1
Green Hairstreak butterfly 162–3
Grimes Graves 29

Hair-grass
    Grey 116, 125, 163
    Wavy 184, 209, 213
Hardy, Thomas 176
Hay Wain, The (Constable) 175
heat stress 16–17, 123
Heath Grasshopper 212
heath hazards 181–4
heath sheep, in German deserts 190–5
    big landscapes and heather 195–200
    heide 193
    heidehonig (heather honey) production 193
    honig 193
Heath Spotted-orchid 145
heather 25, 26, 34, 159, 166, 180, 190–205, 208–9, 211–12, 224
    heather sods 72, 78, 81, 82
Henke, Andrew 177
Holocene 27, 222
Horrebow, Niels 144
Hoskins, W.G. 29, 157, 172
Hounslow Heath 181
human dimension of desert biome 27–32
    Neolithic era 27
    post-Ice Age times 27
hydrology 34
hyper-arid dry lands 13

Iberian badlands 39–63
Iceland 132–55
    black-sand wildernesses 134
    climate of 140
    cold black deserts 132–55
    desert dust impacts 147–8
    desert island 133–5
    environmental history 140–3

flora and fauna 143–7
floral species 154
geomorphological system 139
glaciers 135–7
halting the advance of the deserts 152–5
human-driven erosion 142–3
human history 148–52
Nordic settlers 141
volcanism and dust 137–40
infield-outfield cultivation system 99–100

Jessup, George 168
Jordanian desert 40
Juniper 192, 197, 202, 207–8

Karakum Desert 14
King Lear 176
Kootwijkerzand 65–8
Köppen classification system 16, 44
Kraków 112, 122
Kunming-Montreal Global Biodiversity Framework 220

Ladder Snake 52
On the Land of the Carpathians and Other Polish Mountains and Plains (Staszic) 116
Land of Thor, The (Browne) 150
land-use intensity 1
Large Blue butterfly 116, 117
Les Landes 91–111
    of the agrarian age 110
    agro-pastoral system 98
    Bracken 94
    Dartford Warbler 107
    désert humide 93–5
    ecosystem 108
    fossil-fuel-based substitutes from 108
    from desert to largest human-made woodland 107
    'Great Fire' of 1949 108
    hazards 102–6
    historical Landes farming 96–100
    infield-outfield cultivation system 99–100
    le Sable des Landes (Sand of the Landes) 93, 95
    in the mid-twentieth century 108
    by nineteenth century 106
    nitrogen issue 96
    nutrient stream 98
    plantation 94–5
    Sheep's Sorrel 95
    in twenty-first-century 110
lizards 6, 26, 77, 116–17, 181, 224
Löns, Hermann 204
looming, Błędów desert 126–7
Luc, Jean André de 203
Lüneburg Heide 5, 190–214
    burning 210–12
    farming 190–3

squatter settlement   30, 178
Stanford Training Area (STANTA)   174
Staszic, Stanisław   116
Stevenson, Henry   165
Stone-curlew   167–8, 188
Stour Valley and Dedham Church (Constable)   175
Suffolk Wildlife Trust   6

Tastu, Amable   105
Thesiger, Wilfred   9
Thetford Forest   156, 173
Thomas, Bertram   9
Thomsen, Grímur   149–50
Tiptree Heath   177–9
transhumance   54
'true' desert   12–13
tundra   13, 18, 21

United Kingdom   156–89
    arid Britain   157–8
    See also Breckland

Vatnajökull   136
Veenhuizen village   74
volcanism and dust   137–40
    eruptions   137
    plume areas   139
Voyage en France (Tastu, 1846)   105

Weißflog, Günther   204–5
wetlands   15
Whitlock Rose, E.   106
Wild Boar   52
Wilecki, Stanisław   123
Wistman's Wood   22

INDEX

flora of   206–8
grazing   208
habitat fragmentation   198
habitat mosaic   211
Heather (Ling)   195–9
Heidschnucke sheep   192–3, 195
management   208–13
Nature Reserve   195, 198
visitors   200–5
and World War Two   202–3
Lut Desert   12

Maelifell volcano   151
Mælifellssandur   133
Making of the English Landscape, The (Hoskins)   29
Mangourit, Michel-Ange-Bernard   201
McMurdo Dry Valleys   19, 20
Merlin   146
mesas   24, 44–5
Métivier, Charles-Gabriel-François-Hyacinthe-Denis   105
Montgomery, Bernard   202
Mýrdalssandur   133, 151
myxomatosis   50

Nałkowski, Wacław   116
Nama language   9
Namib coastal desert   9
Natural History of Iceland, The (Horrebow)   144
Netherlands deserts   64–90
   atmospheric nitrogen deposition   87–90
   biodiversity conservation   87
   birdlife   75–7
   distinctive landscape   71
   distinctive species   69
   drift sand areas   71
   drift sands decline and 'reactivation'   83–7
   dry wilderness in a watery country   64–90
   Eurasian Nightjar   75–6
   geology   70
   geomorphological features   89
   grazing   72
   inhabitants   73–5
   inland sand drifts   68–73
   Kootwijkerzand   65–8
   pine plantations   86
   plaggen system   77–80
   plaggenhuts   80–3
New Forest   29, 183
nitrogen deposition   87–90
Norden, John   177
Nordic settlers   141
North European Plain   24, 195, 197, 198, 210, 214

Ódáðahraun desert   137–8
Olkusz   116
Ommerschans village   74

pastoralism   11, 15
Philby, Harry St John   9
plaggen   209–14
plaggen system, Netherlands   77–80, 209–10
   drift sands decline and 'reactivation'   83–7
   harvesting activity   83
   living conditions   80–2
plaggenhuts   80–3
plagioclimax   211, 213, 214
Polish Cochineal species   118–22
Polish Sahara   112–31. *See also* Pustynia Błędowska
populating deserts   61–3, 73–5
Prom-Fiołek, Mieczysław   126
psammophytes   124, 125
Ptarmigan   146
Pustynia Błędowska   112–31
   Diabelskie Pustacie reserve   116
   Olkusz   116
   See also Błędów desert; Polish Cochineal species

rabbits in Breckland   163–9
Reindeer   144
Return of the Native, The (Hardy)   176
Rider Haggard, H.   180
Rochefoucauld, Francois Duc De La   165
Rommel, Erwin   127
Rose, Elise Whitlock   106
Rowan tree   145

Sahara Desert   2, 11–13, 17, 46
saltation (particle transport)   135
Sanchicorrota   59–60
sand drifts and floods, Breckland   169–72
Sand Sedge   162
Sandlings   6, 186
sandstorm impact   148
sandstorms, Błędów desert   126–7
Sargos, Jacques   110
Scots Pine   65, 83, 86, 173
semi-arid dry lands   13
Shakespeare, William   176
Silver-studded Blue butterfly   187
Skeiðarársandur   133
Slow Worm   6, 185
Small Copper butterfly   162, 163
soils   2, 16, 24, 25, 27, 34, 78–9, 96, 141, 157, 159, 222–3
   sandur soils   134
   soil aridity   16
Sonoran Desert   9, 12
Southwell, Thomas   165
Spain   39–63
   Bardena Blanca   41
   Castildetierra   41–2
   La Blanca   41, 57–8
   La Negra   41, 50
   See also Bardenas Reales region